The Autobiogr
China Clay Worker

with a short history of the rise of the
China Clay industry

by

Marshel Arthur, MM
1879-1962

Published by
The Federation of Old Cornwall Societies

The Autobiography of a China Clay Worker

with a short history of the rise of the
China Clay industry

by

Marshel Arthur, MM
1879-1962

First published 1995 by
The Federation of Old Cornwall Societies
from a manuscript written in 1955

ISBN 0 902 660233

Printed by *Swiftprint*
Tel. 01726 70700

ACKNOWLEDGEMENTS

ECC International for photographs of Trethosa China Clay Works

Mr Colin Grigg for drawing of Cornish beam engine and pump

Mr Fred Brent for use of old Foxhole Band photographs

Mr Tim Trevenna (Great Grandson) and his wife Val, for help in preparing the book.

Mr Reg Arthur (Grandson) for supplying the memoirs and helping with production.

Mr Robert Evans publications officer of the Federation of Old Cornwall Societies for his assistance and the Federation for publishing the autobiography

Contents

CHAPTER I

MY CHILDHOOD

In writing the history of a family or country, it is customary to start back in the misty past. In trying to trace the history of my family, I find in the Parish Register that my eighth Grandfather had a son baptised in 1695. The Register only dates from 1694, so before that is conjecture. This ancestor of mine was labelled (gent) and he and his wife were buried "In ye church".

How he lost his land and became a tenant farmer, it is not unreasonable to assume that like the Tregians and Arundells, his ancestors had been "Recusants" who came to this quiet parish for sanctuary. Many of my forebears continued to be farmers, but almost invariably having large families, the younger members branched out into china clay and stone, only farming as a side-line.

So I first saw the light on May 2nd 1879, the twelfth of the thirteen children of William and Louisa Arthur, at a small farm named Middle Hill, in the parish of St Stephen in Brannel, some four miles from St Austell, Cornwall. My father was then a china stone quarry foreman; he also was considered a first-rate miner pitman and dry stone waller.

Wages were not very high, and to help provide for his numerous progeny, he worked in the evenings on his smallholding, keeping a couple of cows, a horse, pigs, geese, ducks, fowls, etc. I cannot remember much about my father but always heard him well spoken of, as a quiet, intelligent, skilled worker, and a practical quarry man of exceptional ability.

He died when I was six, leaving mother with the task of bringing us up. We were happy in having a mother who was a "live wire" in every sense; she was "stir-a-coose" as we say in Cornwall, allowed no grass to grow under her feet, and was boss in her own house - my elder brothers and sisters took no liberties with her. She disliked charity, would much rather work for what she wanted, and of course all of us had to toe the line, and contribute our share.

To show what a resolute woman she was, and how unwilling she was to allow others to do for her what she could do herself, I have learnt since, that when I was born, she never sent for the midwife, but asked a neighbour to hand her the two lengths of unbleached wool and a pair of scissors, passed to our neighbour (Mrs. Hooper). I was washed and dressed, and handed to her in bed, and in a couple of days she was up and on again. What a contrast to these days of pre-natal clinics, nursing homes and analgesia.

FOOD

Now, when everyone has either an electric, gas, oil or coal stove, it is interesting to look back to my boyhood home, where none of these things existed. We had no coal! Our fuel was furze and turf. My father, and later my elder brothers, exercised the ancient rite of turbary, i.e. cutting turf for fuel. The Watch Hill and Longstone Downs are poor to this day because of turf cutting, but what were people to do? - they must have a fire of some sort. So, yearly, pukes of turf were cut and dried, and in the Autumn built into a rick, turfs in the bottom, a couple of hundred faggots of furze on top, then thatched with rushes to keep dry for the Winter .

In our kitchen was a large chimney; I've often sat on a "cricket" (small stool) in it. In Winter there was always a fire, and I've looked up and seen the stars, and many times I've watched mother cooking and baking. First the hearth plate must be just the right heat. To prove this, she brushed back the ashes and sprinkled flour on it: if it browned correctly, all the dust, ashes and hot embers were thoroughly cleared back, then the loaves of bread or cake (cartwheel shape) or pasties, were laid on the plate, and a "baker" (a cast iron dish) turned upside down over them. The ashes and hot embers were raked all around the dish to make it airtight, then a blast of tan-fuzz (armful of furze) was put on top the baker, lit, and allowed to burn furiously at first, then die down into hot ashes; in about an hour the ashes were brushed back, the baker tipped up, and out came the food, baked to a turn.

When I was about nine years old, we had a coal stove installed, but mother, to the end of her days insisted that the "old way" was the best, that a stove dried the food too much.

Though we were "poor people", there was always plenty of plain food - milk, cream, butter, yeast buns and cakes, home-baked bread, pies, roasts, tarts and, of course, pasties. Sunday dinner was the principal meal of the week; this was a boiling. The big crock was hung to the crook in the back of the "chemly". Either a huck of beef or half of a ham, with potatoes in a net, cabbage, turnips, parsnips, carrots in season, suet or chopped tetty pudding, figgy pudding or apple dumpling in their pudding cloths, were all boiled in water fresh from the well, and provided a most appetising meal, served in three courses.

First course: cubes of bread in basins, liquid broth and cabbage dipped out on it, steaming hot, with a long ladle. Second course: beef or ham, suet pudding, potatoes, cabbage, turnip etc. Third course; the fig or currant pudding, or apple dumpling split open and covered with brown sugar and cream.If apples were not in season, it might be treacle dumpling.

In this way we feasted on Sundays, my brothers being home from work, and sometimes my sisters back from service. Mother did her best to make this a satisfying meal.

Another meal I liked was when new potatoes were in season, these boiled in their jackets and served with salt pilchards. Pilchards were bought for one shilling per 100, so we used to split, clean, and salt away in a stug or buzza, about 200. They came out salt as brine, but when boiled, retained just enough salt to offset the freshness of the new potatoes.

But enough about food, or, kind reader, you will be thinking we thought too much about food in the good old days.

However, before I rise up from the table, let me describe just what I can see. I am sitting, in imagination, inside the table by the window, being the south front of the house. At the top and inside the table is a fixed bench, chairs are at the bottom and along the outside; to my left I see the huge chimney, and on the mantelpiece two china dogs, Tom King and Dick Turpin on horseback, and a couple of tea caddies.

Above the mantelpiece a powder flask and shot pouch (the old muzzle-loader gun is hung on the beam); a salt box and pair of bellows one on each side of the mantelpiece.

Next, in the north-west corner is the "Ood Corner", big enough to hold two wheelbarrows of turf, a couple of faggots of furze, occasionally a faggot of "hood" or wood, hence the name. Next along the back wall was the dresser, filled with china-tea and dinner sets, and a wonderful collection of jugs and basins: next, the grandfather clock; on top of its face a ship pitched up and down every time the pendulum swung. At the eastern end of the kitchen was the stairway, and standing back here was the flour hutch and the settle, now put back out of the way, but on cold nights drawn up to the fire to exclude draught.

I have gone around the room, but on looking up, see the bacon rack suspended from the "planchion", which, after pig killing, would be loaded with bacon shoulders and hams; usually a bag of elder blossom, meadow sweet or angelica was also there.

Looking back, that's the picture I see of the old kitchen.

FA 514912

CERTIFIED COPY of an ENTRY of BIRTH.

SUPPLIED AT THE SPECIAL FEE OF 1/- APPLICABLE IN CERTAIN STATUTORY CASES.

THIS CERTIFICATE IS ISSUED FOR THE PURPOSES OF _____ and for NO OTHER USE OR PURPOSES WHATEVER.

Registration District _____

18__ BIRTH in the Sub-District of _____ in the _____

No.	When and where born	Name (if any)	Sex	Name and Surname of Father	Name and Maiden Surname of Mother	Rank or Profession of Father	Signature, Description and Residence of Informant	When Registered	Signature of Registrar	Baptismal Name if added after Registration of Birth
500		Marchel	Boy	William Arthur				June 1899	Registrar	

I hereby certify that the above is a true Copy of an Entry of Birth in a Register Book in my custody.

Witness my hand this _____ day of _____ 19__

CAUTION.—Any person who (1) falsifies any of the particulars on this Certificate or (2) uses it as true, knowing it to be false, is liable to prosecution.

Superintendent Registrar.

[NOTE.—The word "Superintendent" to be struck out and where the Certificate is given by the Registrar.]

Marchel Arthur's Birth Certificate

CHAPTER II

EDUCATION - DAY SCHOOL - EVENING SCHOOL

I started to attend the Board School at Nanpean in the summer of 1885, being then just over six years old. My mother and sisters had taught me the alphabet, and I could read the beginning of the First Chapter of St John. I was in the GO class at Old Pound Sunday School, so after a few days with the Infants I was turned up into Standard I.

I went fairly regularly until the winter of 1889, having just reached Standard V. I loved school, and got along very well with all, passing the examinations yearly, and along with others walking to St Stephen Churchtown to be presented by the Rector with a prize, fully inscribed for passing in Reading, Writing and Arithmetic, signed "Alfred R Taylor, Rector, Wm. Bennetto, Clerk".

I must say that our old schoolmaster, Mr. John Bowden, and teacher, Mr. Elias Allen, drilled the three R's into us. Mr. Bowden could write like copper-plate himself, and I can see the copy-book heading "Be kind to the old man when strong in thy Youth" etc. How we tried to imitate, but how far from the original by the time we reached the bottom of the page. He certainly gave us the "elements", which have served me in good stead up to now.

Of course, I wasn't always a good boy, and he told my brother Jim on one occasion that I wasn't satisfied with education for 2d a week, but had to have the cane thrown in over, but I bear no malice, as I never got more than I deserved, if as much.

Leaving to go to work (which I will relate in the next chapter), I kept on reading everything that I could get. Books were scarce, and there were no lending libraries, yet my neighbours freely lent me from their meagre stock.

After a year or two at work, by listening to men talking, I gradually became aware that I had made a major mistake in leaving school so early. I kept hearing stories like the following. A man just back from Montana USA, said I could have had a slap-up job if only I had had enough education to record the men's time. The manager was very anxious to make me underground captain. Or the story would vary to someone that I knew - he could have been the captain of Yonder Pit instead of contractor as now, and it would have been better for master and men; he's got better ideas and knows how to work a clay-pit the way it should be worked. Really a much better man than the present boss, but alas, through lack of education, he has to work three shifts about, dig hard "for his worms", instead of day shift, using his brains and a pencil.

So I said to myself: "Marshel my lad, if you don't somehow pick up an education, you will be ordered about all your life instead of ordering other people". Result, I went back to Nanpean to what was known as Continuation classes, for three successive winters, studying English and mathematics. Being a Cornishman, English has always been a foreign language to me, and writing or speaking it, I am afraid that I shall always murder the King's English, but I got sufficiently acquainted to make most people understand, but never learnt enough, as grammarians will have noticed, to write correctly.

On the other hand, mathematics had a peculiar fascination for me, so when a teacher was appointed, teaching practical mathematics and elementary engineering, I jumped at the chance, and year after year revelled in maths-arithmetic, algebra, logarithms, geometry, trigonometry and the use of the slide rule, gradually working up to the Binomial theorem, permutations and combinations, and even a few easy equations in the differential and integral calculus.

Being awarded certificates in practical mathematics, heat engine mechanics and hydraulics from the Board of Education and the Union of Education Institutions this qualified me to teach, and I was appointed under Mr. A. J. Edwards as a junior instructor. The four years immediately prior to the 1915/1918 war thoroughly drilled into me as well as my pupils, most of what I'd been taught before, for week by week through these years by lectures, sketches and examples, I drilled them in practical mathematics, using as examples the principle of work, the three orders of levers, the pulley block, the screw, the crab-winch, the principle and output of all lift-pumps, the horse-power of all kinds of engines. So hard did we work, that over 90 per cent of my 18 pupils obtained certificates in the elementary section, and looking around today, I find that all of them, without exception, have found something better than an ordinary labourer's job, so to my young readers I would say, don't be ashamed to take a refresher course, and to other subjects add maths - it enters into everything.

Robert Moreton wrote: "It's the skeleton of God's Plan of the Universe".

Nanpean School

CHAPTER III

WORK - FIRST AT CHINA STONE QUARRY, THEN IN A CLAY PIT

As I intimated in the last chapter, I left the day school in December 1889, and started as tool-boy at Lower Goonamaris China Stone Quarry on January 2nd 1890. Compulsory education up to fourteen years was law from 1770, but was not strictly enforced. My mother being a widow, with no public relief or assistance, my case was "winked at" by the local managers, and my name allowed to lapse from the register.

The quarry foreman was my brother Tom. The wage of 6d. per day was paid monthly (a week kept in hand) at the end of five weeks. To take 12/- to my mother was one of the proudest moments of my life. I remained at the Quarry for fourteen months, having, when I left, 8d. per day. My job was boiling water for "crib" and dinner, carrying and fetching tools from the blacksmith's shop a mile away, chipping china stone, i.e. cleaning the dirty faces with a slad-axe, and assisting in blasting operations, by turning a boryer for a couple of men to hit it. Not a very pleasant job for a boy of eleven, sitting on a damp board with two hammers bouncing up and down a few inches from your nose, and being threatened if you got a three-cornered hole, or didn't keep the tool strictly vertical. Another job I hated was fetching beer from Nanpean or St. Stephens; a two-gallon jar rolling around my back, and I was threatened with dire results if I should happen to let it fall and break it. In one sense it did me good, for I've never cared for the smell of beer since.

How I left the Quarry, was because brother Tom gave me a "blowing up" for being too long fetching the tools from the shop. I resented this very much, because it wasn't all my fault;

11

the blacksmith, having another job on hand, finished it before he started with mine. So that evening I told mother I was "fed-up" and was seeking another job. I soon got one under Captain John Angilley at South Carloggas Pit, then I went to Tom and gave him one day's notice to quit. He flared up again, and said, among other things, "you have been very little good here, only to watch the birds fly over the quarry. Now you are improving a bit, you're leaving." I said, "Well, you shouldn't have blamed me for what I couldn't help" and that was what sent me first to a Clay-work.

The week before the great blizzard of 9th March 1891 found me installed as a tool-boy with a wage now of 10d. per day. On this job, in addition to the usual tools and tea, we had over half a mile of rods from the plunger lift to the waterwheel in the valley; this was part of my job, daily to grease and oil all the working parts. There were four other boys and about a dozen men at this pit.

One of the old customs observed was when a new manor boy was engaged. They "Shoed the Colt". The rule was, on the first pay day, one day's pay was given to the holder, and on the first real wet day a whip around was made, until enough was collected to have a "Join". The tool-boy was sent to buy cocoa, sugar, milk, cream and eggs; this was put in the works pitcher, and, when boiling water was added it was stirred with the hickory stick that the Streamers' boots were packed with; if it was thick enough to cut with a knife, it was considered good cocoa. If anyone refused to pay, they were seized, and the top of their ankle boot was caught in a large 3-crook suspended from the K-beam, and allowed to hang by one leg, head down, until they agreed to pay.

Looking back on those early days, on the whole they were happy years. We worked hard, but we knew it depended on us

12

whether South Carloggas stopped or continued. We thought that our potting clay was the best in the world, and we did our best, under the primitive conditions then prevailing, to produce the pure article. Looking back with knowledge gained later, I've wondered how it was done. We had no cement to point the pits and tanks; it was done with moss. No earthenware pipes, only stone conduits, "conderts" we called them, to send the liquid clay to the pits. Then about 700 tons of the 1,000 tons produced annually was sun and wind-dried; about 300 tons was run across the parish road to a tank at North Carloggas Kiln. I've had the job to watch at Goonabarn, and shout when a horse and cart was coming, to stop the flow and move the launder for the traffic to pass. Leisurely days, etc., but the 700 tons we dried by sun and wind entailed plenty of work.

Shallow pans were prepared, coated with a layer of sand, and a large drop board under the inlet, to prevent the liquid clay from boiling up the sand. Landing operations proceeded, that is, a stream of clay kept pouring into the pan until it was filled up; through three months this would settle to thick clay, as the water was drained off the top. In the Spring and early Summer, after consulting the barometer on a rising glass, word went around "Carren out clay today". All men and boys went back to the "cheeny yard", men with long-handled spades, boys with boards. One man with a cutter would cut the clay into squares, the young men with spades took their own lumps, while one of the older men would lift the blocks or lumps on to the boys' boards, with the sanded part outward, the lads running forth and back setting out the lumps in lines on the ground. For a thin stope, say 6 inches, the price was 1d. per square yard, if it was 9 inches, 1 1/2d, or deeper, 1 3/4d.

I remember earning two shillings one day, when my usual wage was a shilling. I think I earned it; walking forth and back nearly

12 miles, togged up with a stiff knee leather to keep the wet clay from reaching one's chest, and carrying 20 lbs. per journey, was not a bad day's work for a lad of 12.

After the lumps had hardened and were fit to handle by hand, they were put in linhays, or built in hollow piles and covered with reeders. When bone dry, and white as a hound's tooth, women were employed to scrape the sand from the top or sides, being paid a shilling a day, scraping from two to three tons each day.

At 15, being almost man's size, and wanting a rise, I was transferred to Hallew under Capt. Alec Angilley, a brother to Capt. John, to tram sand three shifts about; in the Winter of 1895, a very cold Winter, it stopped by freezing up several of the clay-works.

When Summer came, and water being in short supply, I had a turn at overburden, on contract 2s. 9d. per cubic fathom, which meant to earn 2s. 9d. per day, which we did - 3d. above the normal 2s. 6d., two of us had to break, fill and tram some twenty tram loads a day: good pay then, for a lad of sixteen.

I also had a turn at carrying out clay on a spade at Victoria Yard, near Nanpean: but an opening being available at Goonvean Quarry, where my eldest brother William Henry was foreman, I thought I'd try quarrying again. But working too hard (I wonder), we soon filled up the stock pile, and were transferred to Lower Goonvean to sink a shaft and drive a level necessary to make a "sink". This appealed to me as being in line with the family tradition, for my Father was not only a quarryman, but he and Uncle Will Coon, were considered the best miners for a "wet end" in the district. Shall I digress for a moment to tell a story or two?

Someone had "lost the end" while driving a level at Trethowel, and Mr. R. Varcoe, being manager there as well as at Goonvean, sent them to get things going again, which they did, and doing it as a kind of overtime, they asked the head Captain not to turn their time in until the Summer, as they were realising on their stock pile at the quarry, and they would rather have it later to equalise their pay. And so it was left, but Capt. Will needed a reminder, so Uncle said, "Cappen, you haven't turned in for that job at Trethowel yet." He said, "I'll look it up." Another month went by, still no pay, so Uncle Will approached Cappen again, saying "You haven't turned in anything for that Trethowel job yet." "No," says Capt. Will - who was a religious man - "I've made it a matter of prayer, and the Lord told me that you've been paid." "The Lord never told 'ee that," says Uncle, "the Devil told 'ee that; we ain't been paid."

Skilled Cornish clayworkers were very independent and outspoken. I've heard that my Father, though a quiet man generally, was moved to tell off Captain Bunt in some such words as the following.

"You are a good Cappen. Your office work is all right, a little too sharp on the pencil perhaps, but anyhow, what good would you be at my job? Who sank the shaft? Who drove the level? Who installed the Cornish plunger lift? Who built the pits and micas - standing to this day? You couldn't do it, you know, so stick to your pencil, and don't be afraid to make an extra stroke or two, for there's no harder or more dangerous job than mine in the clay-pit."

Knowing their worth, these men could afford to be independent, for if one firm gave them the sack, they could easily get another job, for miners and pitmen were in great demand before the arrival of the electric motor and centrifugal pump.

But to get back to my working days, we finished the necessary mining, and my brother Tom being hard put for a hand at his quarry, remembering that I left him as a boy - to say the least, abruptly - I agreed to help him. The clayworks boss at Goonvean, Capt. Joshua Key, was very annoyed, saying "You'll go over to that old quarry, and before six months it will "go scat", and I won't give you a job again. I've served you better than most: you're a fool to leave." I said, "Yes, Captain, I know, and I'm grateful, but I've promised my brother."

I went, but it happened just as he had predicted. The Managers fell out and it closed down, but I had the good fortune to get a job as second pitman under my brother Jim, who was working for the West of England China Stone and Clay Company.

At this time (1896) the Cornish Plunger Pump reigned supreme for a fixed pump, and the Drawing Lift, for the temporary pump, and we little thought that it was at the end of a period. Indeed, it looked like the extension of its usefulness, for the clay firms who were deepening their pits were superseding rotary and horizontal engines with draught engines. In 1895 Trethosa, and in the first decade of the 20th century Hallew, Dorothy and Dubbers, installed bigger pumps and engines, 18 inch plungers and 40 inch cylinders. Later, Goonvean installed their engine, which is one of the few working today; all the others have been scrapped. To us it never occurred that after all the good work put in and expense borne by the clay firms, that out of the hundreds of pumps working then, only a few here and there would be left to tell the tale.

I speak with knowledge of the opinion of men and managers about pumps, for during my dozen years as pitman I helped to fix scores around the clay pits and quarries, including the three draught engines and pumping gear at Hallew, Dorothy and

Dubbers. Even when the centrifugal came along at first, we took something like the stand of the old tin-mining Captain when some modern engineers were pulling his leg about his old-fashioned machinery. It got the old chap a bit nettled, and he retorted: "Gentlemen! When you apply the term old-fashioned to a machine, don't forget it's a credit to that machine." Derisively he ended by saying, "None of your new-fashioned 'trade' will ever last long enough to be called old-fashioned!"

He was true enough, because for durability and reliability the old pumps and engines have never been beaten, but in these days, with coal the price it is, the electric motor coupled to the centrifugal pump is more economical. The draught engine, which was considered the best and most friction-free of the old types, when going seven strokes per minute, had to stop and start fourteen times. That fact alone shows the enormous loss of energy, while the centrifugal starts on Monday morning, and doesn't look back before Saturday night. Besides, it is better almost all round; it cuts out deep shafts and long levels, enormously expensive items.

Smaller pumps and pipes formerly took say an 18 inch plus 14 inch plunger to raise 1,000 gallons per minute, now a 7 inch pump through an 8 inch column will do it. It's better for those who have to fix and repair, in the pump house, in from the weather, with a wandering lead for light; whereas formerly down a wet shaft, one had great difficulty in keeping a stinking tallow candle alight. You were beaten sore with the fall of the water on you, and sometimes nearly beaten by the rising water, when effecting repairs. I've stayed more than once until the water got in the seat of my trousers. Once, on a New Year's Day, I remember making a flange joint in water up to my waist, but although cold, it came up gradually, and working hard to beat the water, somehow it wasn't so bad as when you had

17

unknowingly collected half-a-pint in the sleeve of your oiler, and reaching up for a bolt or spanner, it poured down inside your shirt. You'd almost think for a minute that someone or something had bitten you. Beside the discomfort, there was also an element of danger with the old pumps. Two or three incidents will show what I mean.

While changing a plunger pole at Trethosa, 30 fathoms shaft, draught engine (some of my readers will remember the type - massive bob protruding out of the third chamber, from which hung the main rod, having at its bottom end the plunger pole), the first thing to do was to trig the pole "upstroke". This was done by placing a 9 ft. wood trig on the banging beam, asking the engine man to drop the main rod gently until the catch rested on top of this prop. The beam was some 20 ft. down the shaft; it was my job to get the trig lowered and put in place, and to see that it stood quite central, taking the weight of the rod, which together with strapping plates, was about eight tons. This I did, and as we were working against rising water, two men were sent below to unpack the stuffing box and loosen the plate bolts. I had fixed a sheave pulley to the big top bob, rove the winch chain through, and tied it to the newly stocked pole, and was hauling it off the ground to lower it down the shaft, when - slam - went the whole issue down 9 ft. with a mighty bang, followed by cracking of timber and falling stones.

My brother Jim shouted to me, "You never put the trig in vitty." "Aw, iss, I did," I said. "Then how is it gone down stroke?"

That was the problem. With a poor heart, wondering how I should find the poor fellows 180 ft. down below, I jumped on the ladder, lit my candle, and clambered down as far as the banging beam. There I saw that the trig, unable to stand the strain, had buckled out in the middle, and in so doing, had pushed the end of the shaft down, and all below the stays around the main rod, and pump column were littered with broken timber, stones, dirt, etc. It seemed impossible that the two men below could have escaped unhurt, with all the wreckage falling down around them. Carefully threading my way down, making safe on the sets every stick and stone that looked dangerous, getting down nearly halfway I heard someone shouting, "What 'ee left the engine outdoors for? Darn near cut off Dick's fingers!" "Never mind what happened," I shouted, "Come up at once and go out through the adit." That surely was the one time when about a ton was lifted from my shoulders, to think, first, that it was a faulty trig (not my responsibility) and not faulty placing of it; and secondly, that my mates, with the exception of a slight cut on the fingers, were unhurt. What a tremendous relief! With the set-back, it took us until 5.00 a.m. next day to change that pole and secure the shaft, but the thought of what it might have been remained with me all my working life, bidding me take extra care, and I'm happy to say that I never had to report a fatal accident; partly luck as well as good management.

A time or two during my mining and pitwork career, someone nearly had to report me as a casualty. I remember falling into Parkandillack main shaft, into 100 ft. of water, but as I could swim, I got off with a ducking. We were pulling the pitwork out of the shaft, and nearing leave-off time my brother said, "I don't like to leave all that weight on the crab-winch all the night. Take a chain and lash to the pendulum rod." I fixed a plank on top of the pump, some 20 ft. above the water and lashed up, and was kicking the kinks out of the chain as they lowered back and it began to tighten, forgetting as the rod took part of the weight it was pulling my short plank off the set behind me. Suddenly, down I went backwards, vainly trying to grab something: but I fell into the water, getting a wetting, but otherwise unhurt, save for a few scratches.

19

I had another remarkable experience while working - in an emergency - as a miner at North Carloggas. While sinking a cupboard binding trial pit, when down about 60 ft. they cut so much water that a third shift had to be added, so I was borrowed to keep digging all the clock around. Even so, it was a barrel of water to a kibble of muck, which kept the teagle man very busy. An elderly man was on the windlass in my shift; he had lowered down two laths and two end pieces, and he rested while I trimmed out the corners and put this set into position. By the time I had done this, some ten to fifteen minutes, the water had nearly reached the top of my knee-boots, so I shouted up "Lower down the barrel, Harry." But I couldn't get him to move. Getting one end and looking up, I could see him holding tight to the teagle handle, his face the colour of clay. I shouted again and again, and still no movement, he didn't seem to hear me. Then I made a little platform to keep the water (which was rising fast) from going over the top of my boots. I kept shouting at intervals for minutes, it seemed almost hours to me, when looking up I saw him move and rub his hand across his forehead. I shouted again, and believe me, he looked all around wondering where the sound was coming from, before he looked down. Then he asked, "Who are you?" I replied, "You know who I am. Jim Arthur's youngest brother." "No," he said, "I don't know you. Never heard your name before in my life."

It struck me so funny that a man who had been working with me for days didn't know me, that I burst out laughing, moved awkwardly, and fell off my improvised platform, filling both boots with water. Then I enquired, "Can you lower the kibble and pull me up?" "Certainly," he said, "why not?" I said, "All right then, lower away."

It came down, I stepped in with one foot and shouted, "Pull up", but held the other foot ready to catch any irregularities of the timber, in case I made a sudden return.

I sought out the shift-boss and told him what had happened. He said he had had a "fit". "What?" I said. "Is he subject to epilepsy?" "Yes, but only a few know it, for if the Captain found out, he'd be sacked." Still, not quite the man you would like above you; if he'd fallen down on me and no-one came about, we should have been like two drowned rats in a hole.

But all's well that ends well, for one took these things as they came along, and considering the awkward jobs we had sometimes, it's a wonder more accidents didn't happen.

Fifty years ago, to move, say, a 20 ton boiler from Drinnick Wharf to one of the outlying pits was a much bigger contract than it would be today, with modern lorries and vastly improved roads. Back then, a couple of clay wagons were improvised, and from twenty to twenty-five horses hitched in. It was a gala day for the boys and the old horsemen; going around an awkward tight corner with horses plunging, chains breaking, and general confusion, it was pretty near as good as a circus.

Each driver was given a pocket full of split links; after mending up, and a few false starts, on the procession would go again. Next time, perhaps the wheel would sink down to the hub, when out would come the "Jack"; lift up, put planks in under, and on again.

Through these years the rapidly expanding clay trade called for more and more machinery, more pumps, more engines, more drying kilns, more sand tips, the materials for which passed through our hands, so our team grew to a dozen men, known as Jim Arthur's gang, or the "elephant team".

But now I came to a time when I made another move. I had a growing family, and must get all the money possible to bring them up respectably.

An Illustration of the method used by a Cornish Beam Engine
to convey the claystream to the surface from early China Clay Pits

SHAFT

LEVEL

Launder

Cornish beam engine

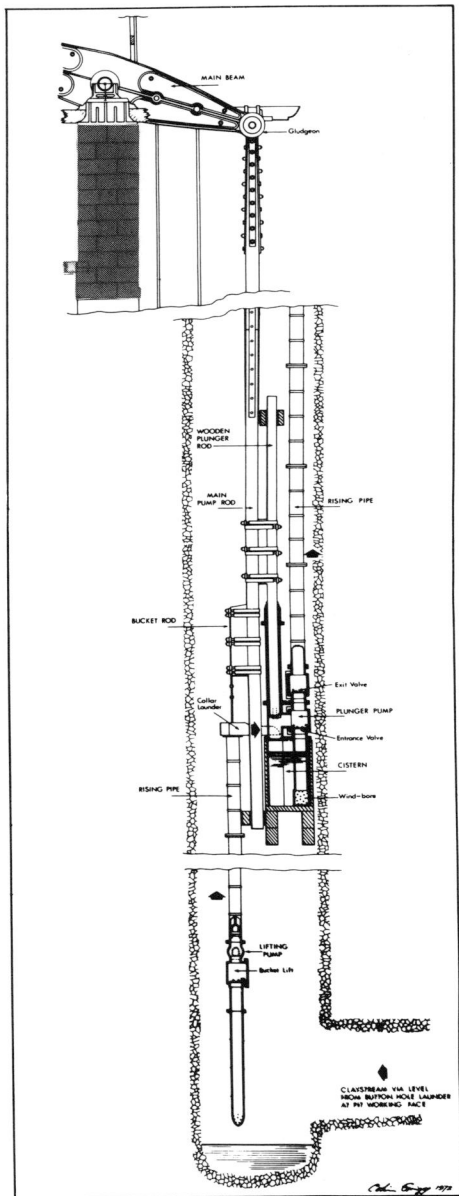

Cornish beam engine

An advertisement in a local paper, asking for applications for the post of Captain at Luke's Clay Work, inspired me to write and apply. Back came a letter advising me to come to their St. Austell office, bringing two recent testimonials with me. I asked Capt. Tom Yelland (the Rounder Boss) for one, and also got references from Capt. H. Jenkin, J. Brewer, J. Grigg and F. Richards. Capt. Tom, when he gave me his, said, "I don't find fault with you trying to better yourself, but we don't want to lose you, Marsh; you are able to take on, should anything happen to Jim." I said that I was happy at my job, but "pay" was a consideration. He said, "I will see what I can do to keep you with our firm", so he telephoned the Managers, and came back with the offer of 3d. a day immediate rise, £5 bonus at Christmas, and the first Captainship available.

A bird in the hand is worth two in the bush, so not being sure of the job applied for, I closed with this offer, and some two years later, in June 1909, Mr. Medland Stocker (a real gentleman) appointed me at North Carloggas.

For the next four years I was happy bringing this small pit into full production, clearing the over-burden back, sinking new shafts, installing new plunger lifts, re-modelling the micas (refineries), re-roofing the drying kiln, and altering the method of sand-filling, so that one man could do more than two did formerly. Then came, in 1913, the great Clay strike, when for eleven weeks the men held off. I remember only two of my men struck the first day - Tommy Wakeham and Josh Hawken. They informed me that a Strike had been called by the Workers' Union; they had no complaint against me, but were simply obeying orders. I said, "It's up to you" - they were good workmen - "it may be beyond my power to give you a job again." They said that they understood, but were being loyal to their Union, so away they went.

Foxhole
St Austell
Sept 21th 1905

This is to testify
that I have known
Marshall Arthur
from his boyhood
and have always
found him honest
and truthfull. and
of a good Christian
Character
Signed John Brewer
Agent of the West of
Eng China Stone quarry
and Society Steward
U.M.F.C

25

Gunnebarn
St Stephens

Gentlemen this is to
Certify – I have known
Marshall Arthur from his
boyhood he have been
working for the west of England
Company nearby all his day
I have found him to be
streght honest Industerious
~~trust~~ man he is well up in
short hand writing as well
I think he will be a very
fitting man for the Purpose

Yours Thos Yellam
Gunnebarn
St Stephens

Gentlemen

This is to certify
that I have known Marshall Arthur
from his boy hood and have
always found him to be honest
and upright in character.
As regards his working ability
he has undertaken several
critical jobs fixing lifts reparing
Shafts fixing launders moving Boiler
Bob & Cylinder for the 40" Beam
Engine that we are fixing here,
and from what I have seen of him
I can recomemend him as being
thoroughly capable of
managing a Clay Work.

Yours Truly H Jenkin
Agent for the W E Co Tubbers

27

Dear Sir

        I do hereby certify
that Mr Marshall Arthur
is a Honest. Truthful Sober
Industrious and Compentent
man and he understand
the working of a Clay Work
and Quarry for he have
worked in the both. And he
is a very good and capable
Pit man and I have know
him all his life time.
And I hope you will esteem
this favour.

        Signed Capt Fred Richards
          East Carloggas
            Works
        St Stephens

I have known Marshall Arthur from infancy; he is quiet, sober, industrious, and intelligent & I have every reason to think he is capable for the position which he seeks; that of Captain at the Gonamanis works.

Signed John Grigg

Capt Hallen Clay Work

22-3-'06

29

Top Row   J Perry   M Angilley   J Merrifield   T Biscombe   D Peters   F Lillicrap

2nd   E Bullen   J Yelland   W A Yelland   E Phillips   J Brewer   M Arthur   J Brewer

3rd   J Neal   H Senitin   W Tonkin   D Bassett   J Agilley   W Light

Some days after, a crowd of men led by Sam Jacobs and Ned Truscott came and called out the remainder, some going very reluctantly and apologising profusely. One I remember almost with tears in his eyes. He said, "I wouldn't care, but I got a fine rick of hay at home, and if I work, someone will catch that afire!" Violence was feared; tempers ran high at times. The Glamorgan City Police were drafted in, and there was one baton charge at Bugle, but on the whole the men were peaceful and law-abiding.

I had one incident. A man came with an order from the Workers' Union Committee that I must stop the pumps and flood the pit. I said that I worked for the West of England Clay Company, and when they ordered me to stop the pumps, I would do so, and not before. "Besides," I argued, "wouldn't it be a foolish thing to flood the pit? You'll be coming back again sometime, I suppose, and it would be wiser to keep the water out, so that you could start at once." He went away, but one afternoon the pumps stopped. I hurried back to the water-wheel, a mile back the valley; when I got nearly there some young men took to their heels and disappeared into the ferns. They were trying to sabotage and disconnect the rods, had taken out the cotter, and tried very hard to knock a pin out of a joint (an impossible task with the weight of the lift hanging on it). I put the cotter in again and started the wheel, and had no more trouble.

Eventually the men came back for the same wage, but a short while after, our firm gave them more than they had asked for, and at 7d. per hour day work, and time and a third for contract, with commodities the reasonable price they were then, many have said, and still say, that, that was the easiest time for a poor man to live, even before or since. Then came the 1914/18 war, but I must have a new chapter for that.

Recruiting drive World War I (Marshel Arthur ringed)

CHAPTER IV

WAR

It's difficult to recapture now the amount of feeling that was aroused in this peace-loving Nation by the wanton over-running of little Belgium, but we soon began to realise that to fight a war, men were needed. Kitchener's posters "Your King and Country need you" were everywhere on the hoardings. I heard Sir Reginald Pole-Carew at Nanpean, and Sir Arthur Quiller-Couch at St. Austell: they made it plain that every man would be needed.

Personally, I was in two minds. My wife was expecting her sixth child in February 1915. I thought it my duty to see her through; unfortunately she was unwell afterwards, and it was not until April that she was well enough to be left. Then I went to a recruiting meeting at Foxhole. A Sgt. Rendle, V.C., was the speaker. He ended his speech by saying, "I've been out once and been wounded. Next week I go again, and I don't mind going to fight for these women and children, but (pointing to us) I'm hanged if I like going to fight for you. You are big enough to fight for yourself." It was true, and I said to my wife, "I can't stand this. I don't want anyone else to fight for me and mine, I'm going to join up", which I did on May 10th 1915.

A number of my men and neighbours signed also, and we were given a week to help in a recruiting drive. I spoke at meetings. Eventually quite a large number, headed by Foxhole Band (several members of which had joined up), marched to the Baptist Church, St. Austell, where Mr. Medland Stocker had provided food for us, then, after some speeches, we entrained for Bodmin Barracks. Nearly all joined the 10th D.C.L.I., but a dozen of us were sent on to Chatham. During the Winter of

1914/15 I had attended ambulance lectures, thinking to join the R.A.M.C., but hearing that it was better pay in the R.E.s and wanting to do as well as I could for my family, I passed a trade test, and joined the Royal Engineers as a Mine Carpenter.

Army life was a revelation to me. Remember I was 36 years of age, had been used to ordering others for years, and had joined up as a patriotic volunteer, but in the depot at St. Mary's one was just another clumsy civilian to be hammered into shape, so we were put through our paces. Shouted at, called "tin soldiers", such phrases as "the other left", "you make me spit blood, you do", and the Irish sergeant saying "Stand properly at attention - put your tums to the teems of your towsers". Having a sense of humour, my friends and I stood up to it, and found that Army life wasn't so bad after all.

The curriculum of the R.E.s was very interesting, to me at any rate. Knotting, splicing, gins, derricks, tripods, bridging, pontooning, revetting, trenching, kept us going, besides drilling and rifle and bayonet practice. At the end of three months we were transferred to Buxton in Derbyshire to be "finished off". We spent our time there digging trenches, night-wiring, pontoon and light bridging on the lake in the Pavilion Gardens, and heavy bridging in one of the vales, route marching up past the Cat and Fiddle, and shooting at the butts in Cheshire, until in December 1915 we were considered fit for overseas.

We left Buxton on December 27th for Plymouth, en route for Middle East, all equipped with sun helmets, but after a few days at St. Budeaux Camels-head Schools, we entrained for Liverpool, went aboard the Olympic, and on January 3rd 1916, sailed for Lemnos in the Grecian Archipelago. Luckily for us, they had decided to evacuate Gallipoli, so we trans-shipped for Alexandria, then travelled overland to Port Said. After an all-night journey in those wretchedly uncomfortable Egyptian

State Railway coaches, we reached Port Said early in the morning, drew tents, and worked all day fixing cook-houses, latrines, and making camp. In the evening I saw on the order board "Corpl. Arthur on 24-hour guard". That was the disadvantage of having a name beginning with "A"; however, there were compensations - I was called first on the pay parade.

EGYPT

Egypt was a very interesting country. On arriving at Alexandria, it was entertaining to hear the vendors shouting in broken English their wares: "Orangeese! Anglese cigaret! Anglese newspaper! All good for the stomach. Very nice, very clean, Johnie!" I liked to hear them chanting as they worked; "Allah, Allah, wyaliz mizraim al Allah", and to see the money changers in the streets, and the old skin water-carrier was like Bible pictures coming to life. In spite of the heat, fleas, lice and sand, I enjoyed it all very much.

After a day or two at Port Said, volunteers were called for to fix a pipeline up the Sinai Desert. As pipe laying was just in our hand, my fellow Cornishmen and I stepped two paces in front of the line, were accepted, and sent at first to El Kantara, where we laid some miles of 6 inch pipeline. After instructing some Welsh Pioneers, I was sent with a detachment to Ballah. Here we laid down some ten miles of 6 inch and 4 inch mains. When this was completed, I was left with a few sappers to filter, pump and patrol the line and its branches. This was a grand job, as I had four reliable engine-men from my own village with me, to drive the semi-diesel and Blackstone oil engines. We took the water from the sweet water canal, forced it through the sand filter, then adding alum through the settling tanks, then under the salt water canal West to East, and up the desert.

We could have passed the "duration" there all right; not but that a junior N.C.O. away from his own officer had his minor worries. I remember the question of where we should camp presented a problem for a while. One evening in my tent, I heard a commanding voice ask, "Who's in charge here?", so I stepped out, and saluted a Colonel. He enquired, "What are you doing here?" Not knowing quite what he meant, I said, "I beg your pardon, Sir!" "Why, darn it, man, you can't camp too close to the natives. It's against all Army regulations." I said that the natives were a bit noisy, but that we had no trouble with them. He said, "You'll have to move." I asked him "Where?" He went down the canal about a quarter of a mile, and said, "You will be all right here", so next day, giving the Arabs a few tins of bully beef, we transported all our gear and tents to this new site, but only a couple of days had passed when a Major, this time, said, "You can't stay here, you are too near that compound of stores; if anything gets stolen your men will have the fault for it," and he wouldn't tell me where to move, so I took boat and went and saw the Area Commandant. I knew him, a very nice fellow, a Captain Grant.

I had done him a favour a short while before. He had come to me saying, "I have a thousand men coming in tomorrow, and the pipeline is broken between here and Ismaila", and he said (mark this), "Will you be so kind as to turn out your men" - it was late evening - "and lay in 400 yards of line, and join up to your system?" "Why, yes, Sir, delighted to oblige." We could work for a gentleman like that. So when he saw me, looking rather downcast, he said, "What's the trouble, Corporal?" I said, "Sir, you'd better get our passage tickets ready. There doesn't seem to be any room for us in Egypt!" I told him my story. He said, "Now, don't worry. I'll send a large boat for your tents and gear, and you pitch on the West side of the canal by the engines, and if anyone interferes with you, send them to me and I will deal with them." We came across, and were happy to stay there indefinitely, but it was not to be.

A few days after, I saw my own Captain from El Kantare: passing him with a salute, he said, "Half a minute, Arthur. I think you may well leave this to Lance-Corpl. Horton. I want you to come to Kantara to run the pile-driver to make a bridge-head for a pontoon bridge, which we are making with some dhows or feluccas commandeered from the natives." I went back, and was engaged for some days with forty natives to pull up the "Monkey" or the 7 cwt. pile driver. We had beaten down a line of 8 inch square pitch-pine piles, and squared them off to take the cap-piece of the bridge-end, when word came - return to Camp! Pack all tool carts: we are leaving via Alexandria for France tonight.

We travelled through the night, reaching Alexandria in the morning, returned tents and tropical equipment to the stores there, got our tool carts and personal gear on the S.S. Scotian, and I remember we had our first lesson on gas-masks before setting sail. After an uneventful voyage to Marseilles, and a 72 hour train journey up through France, we reached Pailleul just as it was getting dark, then a 7 or 8 kilometre walk to Neuve Eglise just across the border in Belgium, after slithering in the darkness over those slippery pave roads, we reached "Tre Roi" Camp, and without supper, and no other reception save the rumble of shell fire, were very glad to stretch out in a tent and go to sleep.

Lucky for us, that was a quiet night. The very next night we had a gas attack, and were called out to "stand-to" with those stifling P.H. helmets on. If it had been the night before, many of us would have been too sleepy to have cared, but, being rested, we got through without harm, although two Australians caught near without their helmets, lost their lives. I saw their lungs in a pail, stained yellow with chloride gas.

After one day's rest, I was despatched with a detachment of carpenters to assemble a sectional hospital near Bailleul. Getting this well under way, I was recalled, made a Sergeant, put in charge of a section, and sent to repair trenches. This started two years on what was known as the Ypres Salient. We did a lot of work in that time, mostly in connection with pipelines, tanks and reservoirs, but taking turns at trench repair, revetting, building machine-gun emplacements, or back a bit from the line, building camps and nissen bow huts by the hundred. One could write a book about events during these years.

I was in charge of a lot of work near Bailleul. My detachment re-roofed all the hangers of No. 1 Squadron Royal Flying Corps. When they built the 'drome in 1914 they covered their hangers with canvas, which, by 1916, had become perished, letting in the rain. So my squad used to rip off the canvas and the small purlines, replace with thicker, and nail on 120 sheets of corrugated iron, and close up each day, not leaving even a nail lying around. We got high praise for quick and careful work, the planes flying in and out all day long.

At the 42nd Squadron we built a new hanger to house six planes, and were engaged in roofing another, when my Captain came and enquired, "Have you nearly finished here?" I said, "Sir, I thought I had, but yesterday an officer from the C.R.E.'s office came and outlined a lot of work, to move the living quarters etc." He barked out, "seems to me you'll never finish here!" I thought, "Hullo, what's biting you?" So when he came next day, instead of speaking to him, I continued hammering on the roof with my men, as good as to say, "If my men aren't doing enough, well, I'll work too." He went away, but next day he was all smiles, and said, "You can leave this to your Corporal. I've got a job for you at Steenwerk, which must be finished by the 28th. You can have all the company carpenters, and I've borrowed four from the pontoon park". Then as a sop, because he had seen that I was annoyed by his remark a couple of days before, he said, "I am raising you to the skilled rate." (from proficient to skilled meant 4d. a day rise)

I went across to this job, and was something of a boss here, with twenty carpenters and one hundred and twenty Argyll and Sutherland Highlanders for fatigue men. My Lieutenant gave me his order book and signed a blank cheque, saying, as it were, "You know more about this than I do." General Schofield used to come on the job and enquire, "Are you getting the materials you require? If you can't at the Company dump, apply at the corps. dump." We did the job with days to spare, and the camp was named after my Captain, "Carter's Camp".

In preparation for the Battle of Messines, operating from Locre, I had the job of getting 4 inch steel pipes as near the enemy as possible. This had to be done at night, so we used to assemble behind an avenue of trees at Hallebast Corner, between La Clytte and Dickebusch, to wait for darkness, then the sentry would let through the lorries or wagons at three-minute intervals. The Germans had this road taped, and one night, the shelling being more severe than usual, I was afraid that I should miss the men of the Wiltshire Regiment, whom I was told to meet at "Confusion Corner" on the way up to Wishaert, a part of the Messines ridge, held strongly by the Germans. So I went through the barrage; as I went along, I came across a direct hit. In the starlight, I saw that the horse was killed, and I turned over the driver, but he was gone also, no sign of life. I went on a bit further, and the shelling getting heavier, alternatively shrapnel and high explosive, I took refuge alongside an old first-aid post.

39

While crouching down there alone at midnight, a rat ran over my feet squeaking. I chased it around the corner there, and saw three dead soldiers laid out. I remember thinking as I looked and dimly beheld a number of crosses where the first aid had buried the dead, of what a returned American miner told me once. When sick in a ghost mining town, on looking at Boot Hill, he said, "I shouldn't like to die here, to be buried over there. Couldn't see or hear anybody." I thought, as the shrapnel cracked and the H.E. crunched, "That's what will happen to me" but the shelling ceased, some soldiers came along, and I said, "Who are you?" The N.C.O. said, We are from the Wiltshires, and are looking for the 167 A.T.R.E." I said, "That's my company - how are you so late?" "The shelling has held us up. There's a terrible mess back there. A salvo fell right on the horse-lines, and killed seventeen horses - rivers of blood running."

The lorries had also got through, and we soon unloaded and camouflaged the sixty pipes the three lorries had brought, and got back to Locre about 3.00 a.m. on June 7th 1917, when the nineteen mines with 957,000 lbs. of explosive went up. It shook the ground, but being spongy country, the shock was cancelled out. I've thought since, if that amount of explosive had been put in one of our deep China stone quarries with something to kick against, it would have blown Cornwall in two.

However, it demoralised and scared the enemy off the Messines ridge, which it was intended to do. All day long the prisoners were marching by us. Our company worked in shifts. I was again on night shift, and in twenty-four hours we had laid down two miles of pipes and a number of tanks from Dickebusch Lake to Wishaert.

I was transferred to Dickebusch in charge of a detachment, to patrol and keep this track intact. For a week all went well, then

Jerry got set, and my word, didn't he pepper that ridge day and night. Hardly a day passed but that the pipeline was burst somewhere, and once four times in twenty-four hours, but I would take half a dozen sappers and repair. We reduced it to a system; with pipes prepared for us at our camp, differing in length by an inch, and the connection by one foot, we were able quickly to breach any gap. Afterwards, to save pipe, we cut up old pipes, and with a special collar of lead wool, effected the repair.

According to our Captain, we were the only company for miles to the right and left of us to keep the water flowing, and I was recommended and awarded the Military Medal. At this time we lived under almost constant shell fire; a strip some half-mile wide , and miles to the right and left of us was pock-marked, shell holes lip to lip, every tree in acres of woods smashed, dead men everywhere. I remember that in digging the trench for the pipeline near a mine at Onraet Farm, they dug through two men, one a German, the other English.

I had one of my boys killed by a direct hit. We sewed him up in a blanket and I got a chaplain to read the service, while I acted as sexton, "ashes to ashes" etc. Another time I felt the wind of a whizz-bang as it sheered over my head and killed a man in front of me. Scenes like these make war hateful to remember.

The Winter of 1917 coming on, I asked my Captain for sand-bags. He supplied us with 2,000 and some corrugated elephant, sections of roofing, to build a substantial dug-out, which we did mostly in our spare time. We tore up rails from an abandoned railway nearby, to make a bursting layer, and shovelled tons of loose earth, besides the filled sandbags. We made double-decker bunks inside, had seats and a table, a door at the end, and even a window with transparent canvas. Our officer said he reckoned it was the best dug-out on the Western Front. But although it was my idea and I worked very hard to complete it, I derived very little benefit, for, about a week after, I was ordered to Kemmel to supervise three catchment areas and pumping plants. I lived in a thatched cottage, and when the shrapnel was bursting at night, one wondered if the hot bullets that we found around some mornings, would set the thatch alight.I went on leave from here in March 1918. While at home, the papers were full of the set-back of the Fourth Army. When I got back to Bailleul, the town was deserted. A town as big as Truro, when I left it sixteen days before it was practically normal. I noticed several wrecked houses as I walked up the Rue-du-Gare to the Ypres Road, and caught a lorry for Locre, to my company.

Next morning I went back to Kemmel, but was recalled after one day. I wondered why, but learnt afterwards that when the Captain came to the Company office, he enquired, "Where's Arthur?" The sergeant major said, "Gone back to Kemmel" and added, "I don't think he's too pleased about it, to be always in the forward lines", which certainly wasn't true. I'd rather be with my own boys, and, in fact, when things were looking shaky, went to the Captain and volunteered to go back there. The sergeant major didn't love me all that much, but he had fallen out over cards with the other sergeant and wanted to do him a bad term.

Things got worse. We were compelled to draw in our outposts; men stopped at the pump houses until machine-gun bullets drove them out. I remember salvaging a Merryweather pump from the Wulverghem Road; then one evening we were told to stand-by to evacuate, G.S. wagons and tool-carts all loaded. We had to take our personal belongings on our backs. I had an extra blanket, jerkin and top-boots, but I dumped everything except my bare kit. So we marched away, skirting Bailleul, into which shells were falling, going through St. Jans, Capel,

Meteren, Fletre, Caestra, to Ecke, where we stopped awhile, moving later to Hondeghem, and began to work on the defence of Hazebroek.

I read somewhere that the British Army at this time dug 5,000 miles of trenches. I can quite believe it, for our Company had 1,000 men attached to it, and each sergeant had 250. Our job was to mark out the trenches with tracing tape, and set the men to work. We made an awful mess of the countryside, but we stopped the Germans, and saved the important rail junction of Hazebroek.

During our retreat (and what a disheartening thing it was to be driven away from ground we had held for years; they called it a "strategical move to the rear", but the fact of the matter was that our part of the line was stripped to stop the rush further south, and the enemy, finding this out, began to push us off the map also), it was bad to see the civilians around Meteren, Fletre and Caestra, fleeing before shell fire, women with their belongings in prams, children clinging to their skirts. I saw a man of sixty years old wheeling his mother in one of these clumsy wheelbarrows; by the pallor of her countenance - a woman over eighty - she hadn't been out of doors for years. An unforgettable, damnable sight, to be helpless, and not able to stop it, made one feel sick.

After some months pipe laying from Zuytpean just under Cassel, to Ouderzeele and Winnuzeele, we were ordered forward again to Rhennigheost. My job was to clean up and dig a storm drain around Dickebusch Lake. I had plenty of men of sorts - the 12th Middlesex, an alien battalion, with Spaniards, Czechs, Poles, Jews, men taken out of the combatant forces. They were all right - in a way - but didn't know one end of a spade from the other. Seeing one place a turf on the spade with his hands, I asked, "What where you in civilian life?" He said, "A ladies hairdresser", and another who had a big voice said, "I shouted the odds, Serg; a bookmaker's tout". I happened to be in Poperinghe, just outside the original Toc H, when one of them, an Austrian (who had been drinking) accosted me. He said, "Sergeant! I don't think the British government serves me fair" I said, "What's your grouse?" He said, "I was born in America, my father had left Austria because he didn't like the Military laws, and as soon as war broke out I joined the British Army to try and smash the Junkers. I was sent to the Dardanelles, got wounded, and while in Egypt recovering, the 'Daily Mail' campaign against aliens began, so they took me from a fighting regiment and sent me to this damn mob. I'm an electrician, yet I'm sent to use a shovel. In fact, you don't know who you have working for you; one man offered last evening £1,000 to start a canteen. In spite of the alien stigma, we are the most patriotic regiment in the British Army".

After setting my men to work, I walked up to Ridge Wood, over a battlefield strewn with scores of bodies, unburied. We were not allowed to touch them, because of booby traps. French, English and German together in death - war's terrible toll. Somebody had buried a Frenchman inside on the Lake margin, so one morning I called for volunteers, a couple to dig him up and plant him again outside, to do it carefully and decently, and when finished they could return to their camp. They did it correctly, and next day, when I stood in front of the 480 men allocating them to their work, these two enquired, "Any more Froggies to dig up, Serg? If so, we would like to be on it", which shows that given an incentive, there's hardly anything but what men will volunteer to do.

After many adventures and narrow escapes, we heard that an Armistice was likely to be signed. We hardly believed it, as shells were still falling around, and had killed the Area Commandant and the French interpreter just the day before, setting alight to almost the last house in Rhenninjhelst, where they lived. I went to have a look, and a couple of tough soldiers were taking out their bones, and saying they must have been a lousy pair, because they hadn't found a watch or a gold Louis, or anything valuable.

But the latrine rumour was true, and on Armistice Day we moved forward - my company to Roncq, 13 kilometres from Lille, I with my detachment to Courtrai. The people were hilarious, the Germans having left the day before. They were sporting ribbons and flags. On the Sunday I went to a service of Thanksgiving for Victory at St. Maerten's Church, at which High Mass was celebrated.

Our job here was to raise a 20,000 gallon tank 30 ft. high, and to join up the pipeline so that the locomotives could take in water, instead of going to Iglemunster. The last act of the German engineers had been to blow down the old tank, and sabotage the pipelines. However, we got along, and made it a going concern, although these were difficult days for N.C.O.s away from their officers. The men were saying - "We joined the Army to fight the Boche, not to build water tanks for the Belgian State Railway. The enemy is beaten, we want to go home!"

However, time passed, and after finishing this job, we were recalled to Roncq for Christmas, 1918. Word was passed around that, if a senior N.C.O. would be responsible, a lorry was available to take a party to see the pantomime "Aladdin" at the Grand Theatre, Lille. I didn't particularly want to go, but the sappers were grumbling that nobody cared for them, so I said, "All right, if you toe the line, I'll be responsible"; so twenty-six of us, including other N.C.O.s who were my juniors, boarded the lorry. When we reached the Grand Theatre, I said, "Now be back soon after 9.30 p.m., this is a walled city, and the gates are shut at 10.00 p.m. If not outside by then, we shall be adrift, and get into trouble." Then we went sight-seeing, and to the pantomime.

Most of the boys drifted back soon after 9.30 p.m.; at 9.45 p.m. three N.C.O.s were still missing. The sappers said, "We're back according to promise; those blinking N.C.O.s will get us in the rattle", so I pulled out just before the gate at the Port du Dunkirk was shut. I didn't report them missing that night, hoping they would have struggled back the 13 kilometres, but as they hadn't turned up, I reported to the Company office.

About 10.00 a.m. I was told that they were in Clink, so I was sent to bring them back under open arrest. When I reached 15, Rue-des-Jardines, Lille, and asked if the three N.C.O.s of the 167 A.T. Company R.E. were there, they looked up their records and said, "Yes". I said, "I've come to take them back to the Company". I accompanied the warder down to their underground cell, and looking in, saw Bert and Tom head in hands, while Jock was playing patience with the raggedest pack of cards I ever saw.

I said, "Hullo, boys! What are you doing here?" They jumped up, saying, "Have you come for us?" I led them to a restaurant for a coffee and bun, and on the way back to camp learnt that vin blanc (white wine) had led to their downfall. On entering the Grand Theatre, where smoking was forbidden, they refused to put out their cigarettes, and were man-handled by the Red Caps. They struck back, and were arrested. They were rather crestfallen, enquiring if I thought it as disgraceful to be in a military as in a civilian prison?

Bert said, "What do you think? After a wretchedly cold night with one blanket on a bare board, the warder came in with an apology of a breakfast. He said, "Hullo, Bert, what you doing here?" - a man from my own town. Next month I'm due to go home and marry a very nice girl, now he'll write home, and what will her people say?"

Tom said, "That chap in the Company will write to Stourbridge, and my wife will know of this", but Jock was imperturbable. Nobody from Greenock knew about him, and he helped to get a light sentence, for when before the C.R.E., who said, "I'm sorry to have you men before me on such a serious charge, having gone through the war without a mark against your character. How were you so foolish?" Jock said, "Well, Sorr, we were a wee bit merry, you know." The Colonel could hardly help smiling, and said, "I'm letting you off light. You'll have seven days pay stopped, but your papers won't be endorsed."

Being myself a teetotaller and non-smoker, I could never see what there was in "drink" that made men so silly, and get into so much trouble to indulge in it. An Irishman, by far the best soldier in our Company, twice reached full corporal and back again. When in Egypt he stabbed a man, and got second field punishment, three months without pay. We had whip-arounds for him. The last I heard of him he was cook's mate at the base; he might have been a colonel, but for the drink - he had ability.

Old Tom, again, was one of the best-liked men in every way (but one), an admirable old Yorkshireman, but what he would do for a drink! Set to clean out the Captain's office, he found and drank his whiskey, and had C.B. for it. Another time he fell into a trench of water, and came into my hut at Neuve Eglise, drenched to the skin. We stripped and towelled him, one giving him a shirt, others pants and trousers; we rigged him up. Was he perturbed? Not a little bit - he was singing "Down went Maloney to the bottom of the sea, we haven't got him yet, he's bound to be very wet". Tom had a bosom pal, McNalty from Glasgow, who, when fuddled got run over with a lorry, and was killed. We lost Tom one pay night, and he was found lying

Marshel Arthur in uniform 1918

beside McNalty's grave. When questioned as to why he got there, he replied that he thought McNalty might be lonely.

Somehow you couldn't help liking a man like that, but what a pity that drink should spoil some of the best of men! But I am moralising.

In February, 1919, I was released from the Army, being a pivotal man in the clay industry, and so returned, after three years and nine months, to civilian life again.

Trethosa, showing old beam engine and kilns in background

CHAPTER V

STARTED CLAY WORKING AGAIN - SLUMP - TRANSFER TO NEW PIT - SPECIAL WORK - RETIREMENT

On returning, and being reinstated at North Carloggas, I found that day work pay had risen from 7d. to 1s. 6d. per hour, though it was doubtful if it had kept pace with the price of commodities. However, as is commonly known, times got worse and worse, until pay got back to 1s. per hour. The clay firms did their best in the very difficult situation that had arisen; they employed the married men, and organised a lot of improvements - unproductive work.

I was in charge of some, such as relaying the Dubbers Burngullow pipeline past Carpalla Work, building a new saw-mill at Drinnick, and a new line and shed at Quarry Close for Hendra clay pit, besides getting the clay required at North Carloggas.

In January, 1925, I was transferred to Trethosa, when Capt. Daniel Bassett retired. The war years and the situation of the work, hemmed in nearly all around, made it imperative that something drastic be done. The pits and micas, engine house and railway line were in the way, stopping development. We first tackled the pits and micas, as one pit had burst its walls and fallen, clay and all, into the bottoms. A site was chosen the other side of the parish road, and pits and refinery capable of dealing with twice the former stream, were built. This was a rather big job, taking two years to complete; there was an immense amount of mason work, with over a million cubic yards of "backing", which we hauled out of the adjoining Kernick Pit.

Fortunately, I had in Capt. Stan Fugler (master mason) and Foreman Jack Biscombe, general labour Captain, two colleagues who loyally and faithfully carried out the plans which I had drawn.

Since finished, over one-third of a million tons of Class A clay have been refined, settled, and sent forward to the kilns. Now, 25 years after, other methods are coming into vogue, but for many years this refinery was considered A.1. Lord Aberconway once said to me, "I congratulate you, Captain, on your 'lay-out'. It's the best in the china clay area."

Our next job, as the old pumping engine looked like falling into the "bottoms", was building a new house and installing a suction gas plant, for generating electricity; then we put in a centrifugal pump, dismantling the old engine and Cornish pump.

Meanwhile the firm had built a long drying kiln at Tolbenny, capable of drying as much as the three kilns we had at Trethosa, so we were able to move the railway and develop the pit by opening out on top. Soon after the advent of the suction gas plant, we started to "sink", and went down 70 ft. with a gravel pump. Finding electricity giving efficient service, and being easy to install, English Clays Lovering Pochin built a power station at Drinnick, and a sub-station at Trethosa, so during my tenure there, we changed from the draught engine to suction gas, and then to the sub-station with power on tap, with, I'm bound to say, improvement every time for the engine driver, who formerly had to get in an hour earlier to get up steam. How much easier to put in the switches, and away.

During my working life, alterations and improvements have been brought into use in almost every department of china clay production.

Dealing with sand: from ground pit via patent pit and flying tip, later the scrape. Burden: instead of men breaking, filling and tramming, it's now the excavator or scrape. Pumping: now almost universally the centrifugal instead of the Cornish plunger pump. Engines: electrically driven, instead of steam. Washing: power hose, instead of the manual breaker and washer. Drying: old styles sun-dried, later clay trammed from tanks and spread over the pan and cut into cubes, and now the modern filter press and cake.

When I was promoted to "Captain" in 1909 we sent one sample each week to Head Office (it then took about four months to condition clay for economical drying), and we took stock twice a year.

When I left, on some clays we took a sample every three minutes. During a rush, clay was taken from its virgin bed, washed, pressed, dried, and put on ship at Fowey in forty-eight hours, and we took stock every week - fifty more judgement days each year than formerly for the poor old Captain. So china clay producing has been speeded up to meet modern conditions.

During the 1939/45 war we kept going. It was not easy, with men leaving and shortages of materials, but with able management from our Heads, we won through, and wages reached 2s. 6d. an hour.

In the tough Winter of 1945/46, following an operation, I had a nasty bout of 'flu so, not feeling quite equal to carrying on - ours was a deep pit, some 360 steps to climb out of - I thought I'd better give over to a younger man, and I asked to be put on the retired list on September 30th, 1947.

My men gave me a clock, duly inscribed, and the following is what appeared in the "St. Austell Guardian."

Trethosa showing monitor at work

Trethosa, view across pit showing inclines and tips

BROADCASTING CAPTAIN RETIRES
WORKS PRESENTATION TO MR. MARSHEL ARTHUR

On retiring as Captain of Trethosa China Clay works, Mr. Marshel Arthur of Foxhole, still robust and well at the age of 69, on Saturday received a presentation from the employees of Trethosa Works, and tributes were paid to his long and devoted service in the china clay industry in the St. Austell area.

Mr. Percy Dyer (Manager), introducing Mr. William Pascoe, who made the presentation, said he had worked with Mr. Arthur for 20 years, and he knew that a Captain did not have an easy job, because he had to try to please everyone. They all wished Mr. and Mrs. Arthur every happiness.

Mr. Pascoe, who has the longest record of service at Trethosa Clayworks, handed to Mr. Arthur a clock inscribed from the workmen of Trethosa Pit.

Mr. Arthur, in acknowledgement, said he had had a very long innings in the China Clay industry, in which much progress had been made; it used to savour of brute force and ignorance, but now it's the age of the machine. "If you can take a burden off a man's shoulders and place it on the unfeeling iron or steel, it's all to the good, and something worth while." He was thankful for all the help he had received from Managers and men.

Mr. Arthur, who had worked for 57 years in the China Clay and Chine Stone Quarrying industry, comes of a family which has lived in the china clay bearing district for eight generations.

There is a family association with William Cookworthy, the discoverer of china clay in Cornwall, who, in St. Stephen-in-

Brannel parish nearly 200 years ago, found deposits of china clay which he could use in the manufacture of porcelain. Cookworthy, indeed, stayed with Mr. Arthur's great-great-grandfather at Carloggas Farm, which is in some ways the birthplace of the china clay industry.

First employed as a second pitman by the West of England Clay Company, Mr. Arthur became foreman at North Carloggas in 1909, remaining, except for the war years, until 1925, when he was appointed Captain at Trethosa.

He served during the First World War in Egypt, France and Belgium, as a sergeant in the Royal Engineers, and was awarded the Military Medal. He did Civil Defence work in the last War, and qualified for the Defence Medal.

Mr. Arthur has broadcast once with Mr. S. P. B. Mais, the author, and the late Mr. Alfred Davies, and once during the War, to the Middle East. He was in the film "Cornish crafts and Industries", and over the pseudonym "C.C.C," (Cornish Clay Captain), he has contributed articles to the China Clay Review.

In his retirement, Mr. Arthur is contemplating writing a book on his life and experiences.

Mr. Arthur, who started work at 10 ½ has been concerned in the invention and introduction of some of the mechanical equipment now in general use in the clay industry. With the aid of his brother, Capt. Jim Arthur, and brother-in-law, Capt. H. Jenkin, he helped to plan and erect the first flying tip at Dubbers. He drew the sketch of the first V-shaped wagon which made the flying tip really practical. He also worked out the principle of the Hydro-vac, a system of sand and mica induction, which has been of value in the industry.

Mr. Arthur is a Methodist local preacher, and an active officer of the Foxhole Methodist Church. He was also a bandsman in Foxhole Silver Band, for more than 30 years.

CHAPTER VI

WIFE AND FAMILY - PATRIOTIC SERVICE

Away back in Chapter II I said that at sixteen years old I was earning a man's pay. I must have thought I was grown-up, for I started courting. After four years "walking out" I had a house offered at a low rent, £3 per annum, so we decided to marry.

My brother Jim told me, "By right you should have at least £200 before you start in double harness." I was earning under £45 a year; if I saved half of what I earned, it would take me eight or nine years to get that amount. Mind, my brother practised what he preached. He courted for seventeen years, and by raising stock and working overtime, probably had £500 when he got married. I had about £20, and when you are twenty you are not afraid of risks, so we tied the knot with our tongues which we have never been able to untie with our teeth, and after fifty years together we both think we could have gone further and fared worse.

We've had seven children; one died at birth, the other six grew to men and women, all happily married. Three sons, three daughters, eleven grandchildren, two great-grandchildren, is our grand total at seventy.

I suppose that we can claim to be a patriotic family: myself in the 1914/18 War, my eldest son (poor fellow) at eighteen joined the Navy in the First World War, and kept on in the Reserve. He was called up in 1939, and put on the Western Approaches; later, in 1940, on the Dover patrol, he went into Boulogne harbour to take off some Service Unit, and all the superstructure of the Destroyer Venetia was blown away by direct gunfire. The crew suffered twenty-six casualties; they got back to

Devonport, were repaired, went back again, and got caught by a vibration mine near Gravesend. The Venetia was blown in two and sunk, and my son was in the water an hour and a half before being picked up. His next ship was the Hecla; while going to South Africa it was mined. Repairs were made at Simonstown, but on the way back to North Africa landing in 1942, just outside Casablanca, the ship received three torpedoes and was sunk. Our son was one of the 200 odd reported missing. This was a terrible blow to us, especially to my dear wife, as well as to his wife and children. I fear that they will never quite get over it. He left four children, two girls. One married a soldier, the other an airman. His elder son served with distinction in the R.A.F., leaving with the rank of Warrant Officer, and was awarded the D.F.M. for thirty-five raids in a Lancaster over enemy territory. The younger boy was at school.

My second son was directed to work as a carpenter at St. Eval Aerodrome, later as ship's carpenter at Malpas and Charlestown. Of his four children, three boys joined the Navy, two as Ordnance Artificers and one as E.R.A., while the girl served as a nurse in the Military Hospital at Bodmin, later moving to Chester City Hospital, where she gained her S.R.N. Certificate.

My youngest son served with the Royal Army Ordnance Corps in North Africa and Italy.

Two sons-in-law joined the R.A.F., one serving in India, the other in the Bahamas. The other worked at Filton, Bristol, on planes, being also a sergeant in a Home Guard bomb disposal squad.

My second son and myself served all through the war as part-time air-raid wardens, as well as being borrowed by the Home Guard for cliff demolition. My eldest daughter still serves as a part time ambulance and Red Cross worker. No doubt many other families have done and suffered more, but at least we contributed a little in our Country's dire need.

What advice can I give (who will take it, anyway?) on marrying early and raising a family? Some decline the responsibility, but to one childless man who used to jeer at the folly of having children, I said that a man without children was like a figure 9 with its tail cut off, exactly equalling 0.

To anybody with a small wage it's a mighty struggle to raise a family. I was warned it meant twenty-one years in the gutter; seven years going down, seven years down there flat, and seven climbing out again. But there are compensations. To see in later life your children rally round in sickness, or any emergency, or indeed at any time, is worth all the years of struggle.

I was fortunate in picking the right partner, a good manager, able to cook, a splendid washer. We never got into debt - what we couldn't pay for we stayed without - and she could make and mend for the children. She was never idle - not a lazy bone in her body - and at seventy she is still the same.

Dear reader, you can please yourself about marrying early, but if I had to live over again, I would do the same.

Standing: Marshel, Louie, Kate (mother), Lilian, Ralph

HMS "Venetia" - showing damage caused by abandoned French coastal guns when entering Boulogne harbour to evacuate British troops 1940.

63

Top Row   *Medals of Marshel Arthur, MM, Royal Engineers 1914 -1918 War*

2nd Row   *Medals of Launcelot Arthur, Royal Navy, 1914-1918 War and 1939-1945 War*

Bottom row   *Medals of Reginald James Arthur, DFM, Royal Air Force 1939-1945 War*

CHAPTER VII

SPORT AND SPARE TIME HOBBIES

Organised sport, as we know it today, sixty years ago had not reached our village; we had no football or cricket club, or playing field. We assembled on a green plot in Jan Bullock's dam park, to play "rounders" once a year, on Good Friday morning. Sometimes in the evenings we played a rather rough game which we called "hurlo" (something after the style of St. Columb hurling), but marbles (small ring) was the best-liked game; grown men, as well as boys, played for hours in season.

As playing fields came into fashion, I helped, with others, to level the ground at Stennack Wyn (formerly worked for tin), to make a playing field for Foxhole, and I also helped to raise money, and with other volunteers, to build the pavilion.

At Nanpean, the Playing Fields chairman, Mr. James Neal, and the committee, asked me to do the levelling and surveying to lay out the Victoria Bottoms Playing Field. With my three sons, it took quite a lot of measuring and calculating to set the bench mark at what height the football pitch should be, so that all the irregularities should be ironed out, having enough, but not too much soil, to get the field with correct drainage. Voluntarily, the Nanpean men worked very hard; it was a pleasure to co-operate with them. The result, as may be seen today, is very satisfactory, time and again it has been quoted as being one of the best pitches in Cornwall.

But my spare time hobby was being a member of the village band for fifty years (on and off), sometimes in the reserve called up for contests. I started in 1896, played at Queen Victoria's Diamond Jubilee in 1897, and was playing through the last war up to 1946. Music is a wonderful hobby. I never was a first-class musician - a player of minor parts; as a Band we had a measure of success through the years. At first it was a small band of from nine to a dozen - the following is a bit of my doggerel describing them in the first decade of the 20th century.

DOGGEREL

I think you'll agree, for 1903
The Foxhole Band were absolutely grand.
In years past and gone, they were dubbed A.1
In that year of grace, they had reached first place;
On their fine cornet end, they could ever depend,
They could blow high or low, or play music fast or slow.

Bandmaster Will, blew clear and shrill,
Lieutenant Mart equally smart,
Johnnie rather Young, but coming to come
Wakeham of the right sort, a blazer for double fort.

Tenors Marsh, Dave and Ob between them do a fair job,
Marsh a modest brother, deputise another,
Dave in hats and ties is up to date,
And helps to make up for his mate,
While Ob with the tenor trombone
Rated really second to none.
Euphonium by Will is played with great skill.

But although the smartest of the smart
Is usually quickest to lose heart,
His splendid musical abilities
Cancel eccentric idiosyncrasies.

Arthur with the E flat bombalore,
Dave with the B flat make two more.

These are really smart young fellows,
With wind like a blacksmith's bellows.

Last but not least is Mr. Hawke,
Who with drum keeps time like a clock.
When after marching through town or village,
They give a select programme on the stage,
Music delightful! Exquisite! Unique!
The kind that makes everybody speak -
Some in tones of praise, others their voices raise.

What they say don't amount to much
Every member class all such,
As ignorance, envy or lies,
And are quite perfect in their own eyes.

"One and all", that's how we used to stand.
A truly "Vitty", Cousin Jack Band.

With the years, the Band grew, and started competing at contests, obtaining at least one first before the 1914/18 War. Many of the bandsmen volunteered, and played in Army bands. On demobilisation in 1919, the Band reformed, and because the E flat bombalore wasn't released from the Army, I took on until he came back. We started contesting, and won three firsts at St. Dennis, Newquay and Bugle.

1920 brought back the bass player, so I took a tenor horn just to fill in, and then as the solo horn player just before a contest broke his leg. I was promoted to solo horn because the first tenor wouldn't move up. That year we won six firsts, making nine firsts in succession, ending up at Bugle again.

In 1921 the band was short of cornets, and only a schoolboy on solo, so I was again moved, to take a cornet. Through this year, other Bands had caught up, and we could only reach second prize, but having our old trainer, Mr. Joseph Stubbs, of Crewe, down for Bugle, we just won by three points, over bands that had been beating us at other contests all the year. By doing this, we won the Sir Edward Nichol cup outright, and each member became entitled to an enamel and gold medal with his initials engraved on it. Sir Edward came to Foxhole with Mr. F. J. P. Richards, Hon. Secretary of the Bugle Contest, to present the medals personally.

When it came to my turn, in his bluff, hearty way he said, "Well, old chap, what instrument did you play?" I replied, "It's singular, you asking me that. First year E flat Bass; second year Tenor Horn; third year, Cornet. If I didn't do much to help, I couldn't have blown many wrong notes, could I, Sir?" "No, old chap," was his answer, "you ought to have three medals!"

Looking back through the years, one sees how contesting has raised the standard of band music in Cornwall. Fifty years ago, I'm afraid most of us thought noise was music . Our local Bandmaster tried his hardest, but many of us were like the comic member of the Band when we were on the way to a town where the people were musical, and the Bandmaster was cautioning us, saying, "Expression is the soul of music; whatever you do today, mind your P's and Q's - do'ee mind your expression". The comic one said, "Dear, dear, I came away and properly forgot mine! Let it home on the piano!" I am afraid that often enough we were all like that.

Anyhow, the Midland trainers, who came down for a week before a contest, used to let us have it. I can remember Mr. Stubbs, a fine figure of a man, six feet tall, big in proportion, smartly dressed, with pointed moustachios, shouting at us for mis-accenting some syncopated bars - "Whatever did you do

Foxhole Silver Prize Band about 1898
(with Captain Marsh Arthur)

Foxhole Silver Prize Band about 1910
(with Captain Marsh Arthur)

Foxhole Silver Prize Band about 1922
(with Captain Marsh Arthur)

that for? It's as "wrongh" as wrongh can be! If I had a pistol, I'd shoot you." Then he calmed down, and said, "There's nothing to beat trying, but doing," and over and over again we went through it.

And so, by sticking intensive training, bandsmen began to see that there's a lot more in music than noise. Great praise is due to local organisers, such as at Bugle, Stenalees and other places, for year by year doing so much voluntary work to raise the standard of music.

CHAPTER VIII

PUBLIC SERVICE, RELIGION

In recording one's activities, please don't think that I am boasting. I have never felt a very able public man, not having - as many I've known - a flair for it.

I was co-opted a wartime Parish Councillor, and made a School Manager, serving the usual three-year term.

When a Parish Invasion Committee was elected, my job was Water Supply. Now I represent Foxhole on the St. Austell Rural District Council.

I was Hon. Secretary of the local branch of the British Legion for some years, also Chairman for a time.

As A.R.P. warden starting in 1938 and continuing until the end of hostilities, it was quite a task fitting the inhabitants of our large village with gas-masks, attending lectures, and being on duty in air raids. We had fifteen bombs dropped one evening at the lower end of our village, damaging some forty houses, and killing four people. Many nights we spent on duty after the sirens wailed. Then the wardens, with the W.V.S., dealt with evacuees, receiving, giving them a meal, and conveying them to their billets.

My wife and I had at various times, two boys, two girls, a teacher and a mother in our home.

I have not a long record of any secular public work, but have helped a little here and there to the best of my limited ability. My longest and most continuous service has been for village

Methodism. I joined the Foxhole Methodist Church in 1896, and "got converted" in a Mission led by the Rev. John Whale, father of Dr. Whale, author and B.B.C. broadcaster. That led me from being a rather wild, irresponsible young fellow, to take an interest in organised religion. I joined the Sunday School and became a teacher; I am still a superintendent after fifty-four years.

At the Sunday School Jubilee in 1930, I was presented, along with others, with a silver medal for over thirty years' service. I also became a pledged abstainer, joining the Band of Hope, at the monthly meetings taking part in recitations and dialogues. In this way, I got over "stage fright" and "platform fever". I think it would be better for the youth of today if they made some of their own entertainment, instead of buying it all at the cinema.

As Society Steward and Hon. Secretary of the Trustees, I have tried to help my own church in all its varied activities, presiding at hundreds of miscellaneous gatherings. During the war (1940) our superintendent Minister, being short of preachers, put my name on the Plan (without asking me), first as a helper. Later I was promoted to full Plan, so for ten years I have, in a humble way, tried to preach the Gospel.

Whatever, through the years, I've done for my Church, I would like to place on record that I've received much more than I've given. The fellowship of the splendid men and women - many of them gone to their reward - and the comradeship, inspiration and example of their Christian lives, have steadied and helped me in times of stress and strain, and - under God - I shall be eternally in their debt.

CHAPTER IX

IMPROVEMENTS AND INVENTIONS

As I remarked in a former chapter, great improvements in all departments of the production of china clay have come about during my working life, and I, personally, have been happy to assist in pioneering some of these new methods.

Take the improvement in handling sand. In the production of china clay, to obtain one ton of clay about four or five tons of sand have to be dumped. Anyone passing through Mid-Cornwall can see these conspicuous conical hills, made up of millions of tons of dumped sand, evidence of the immense amount of work done to get china clay.

The oldest method of dealing with sand, I've heard old clay-workers say, was to "land it", i.e. shovel it out of the running clay stream on to the bank alongside, then wheel it away in wheelbarrows. In my boyhood, the ground pit was used. Men with long-handled shovels filled a tram wagon from ground level (hence the name), which meant lifting the heavy sand, an enormous amount of work.

To avoid this, at some works the incline skip-wagon was run back underneath the pit. This was labour saving, but somewhat difficult and dangerous - difficult to keep the channel clean, and dangerous pushing back the skip, if the engine-man mistook two for three, and "pulled up" instead of lowering. Fatal accidents have occurred with this method.

PATENT PIT

So the patent pit came into being. While working at Dubbers Clay Mine in 1907, my brother-in-law Capt. Harry Jenkin asked me to inspect with him "skip-wagon under-pit method", as he wanted something speedier than the old ground pit. We visited a clay-pit, and saw it in action, and turned it down as being too slow and dangerous. We designed and built pits for a hand-tram to go under, the wagon was filled by removing "traps" and digging sand down into it; when full it was trammed to the "dogshole" and tipped into the skip.

In an article in the "China Clay Trade Review" of April 1922, I claimed for this method the following advantages. I wrote that it was easy to construct; ensures dry sand; wagon can be filled while the skip goes up and down; being away from "dogshole", there is no danger; it is on right principle of work - gravitation for not against, as in the ground pit, consequently there is ease and speed in filling wagon; the result is that a large volume is moved in a short time, which makes for cheap labour - low price per cubic fathom.

I made sketches of the method of framing the tunnel and fore-end of the patent pits, as we called them, for many of the clay works under our firm, and now, forty years after, in many places they are still in use.

It is true that some pits have mechanised the handling of sand; there can be no standing still; we live in an age of progress - a mechanical age.

But any system that held good for forty years must have had some merit, and even today, with a battery of well-designed pits to ensure dry sand, it is easier to keep skip and incline clean than with some of the all-mechanical layouts.

DERRICK TIPS

I helped as No. 2 in Jim Arthur's Gang, together with Ganger Allen's tram-line party, to raise the first derrick tip at Dubbers Sand Dump. "Imitation" someone wrote, "is the sincerest form of flattery", if so, I for one should feel flattered, as I look around on the sand and burden dumps of today, for they are invariably crowned with a derrick.

It is true that there are variations in application and in methods of "pulling forth", but we laid down the main principle of suspension, which still holds good, and will do so as long as skip-wagons run.

In describing these tips in the "China Clay Trade Review" of December 1920, as an advance on the old-fashioned way of tramming the sand from the poppet-head by hand tram, this is what I wrote:-

"Just a word about the derrick or flying tip: these are very noticeable in the china clay district, on the top of the conical white mountains near the edges of the clay mines; as can be seen, it is a kind of derrick with wire ropes, guys, etc. to keep the end of the incline road up, so that the skip-wagon can have a clear tip. The advantages of this tip can be summed up as follows:-

first, it requires no hand tram on bank top; second, no rails or sleepers; third, no wood trestles to raise the poppet head; fourth, it keeps the sand in the smallest space possible; fifth, tip does not require adjusting, sometimes for a month; sixth, one lad can deal with any amount up to 200 loads in six hours; seventh, sometimes with automatic tipper and shutter, and when engine-man can see the tip top, no one is required at all."

Trestle, an incline and tip

THE SKIP WAGON

The old-fashioned skip wagon butt was rhomboidal in section, swung in the frame on two hang-bow-pins which revolved in two pedestals when the tipper was released. The tipper was a piece of hardwood 4 inches square and about 3 feet in length; it was released to allow the butt to tip, by a blow from the tip boy's wooden mallet, a clumsy, awkward arrangement. Sometimes the boys could hardly get it back into position again, if sand stuck in the lip. Something automatic and speedier was urgently needed, and so the automatic V-skip was invented, to work in conjunction with the derrick tip.

V-SHAPED SKIP

The following is what I wrote in the 1925 June number of "Chemical Age" (China Clay supplement) about changing shape of skip-wagons.

"If there is any blame or fame for helping to bring this about, I plead guilty to being one of the pioneers, as I drew the plan for the first V-shaped skip, although the idea came from my brother. It came about this way.

I had designed an automatic discharging skip, retaining the old shape, but instead of swinging up-side-down to tip, I built the butt on the frame, and had a hinged steel plate bottom, which, dropping down when the catch was released, instantly discharged the load. On the return journey, the swing door came into contact with a piece of wood suitably bevelled, and was lifted back into position to shut off automatically. This worked well for days, but the wagon getting off the track bent the steel plate, which then refused to function properly, dropping loads on the incline, and it was thrown out as one of those very good inventions that wouldn't act.

Being younger then, I was naturally disheartened, and almost vowed to leave skip-wagons alone for the future. But my brother said that my idea was sound. He said it was much better to have a door to open than a skip to turn upside-down, as the skip could be built stronger, and the load discharged with less clearance. Where I failed, he went on to say, was in placing the door at the bottom, where, if it accidentally unfastened, or the skip got off the track, it would get damaged and fail to work. "But, my dear man," I said, "where else can one put a door but at the bottom? The load won't fall out if I put it on the side!" "Yes, it will," he said, "only the butt must be triangular in section, instead of rhomboidal".

So I drew the plan, and the local blacksmith and carpenter made it. Many improvements have been added (by others) since then. As I've written before, those of us who have helped in any way to make this skip the undoubted success that it is, do not claim for it perfection, but as the centre of that remarkable quintette - Patent Pit, Expert Sandmen, Skip, Flying Tip and powerful well-handled winder, it is something very near it anyway. It is the fastest and most economical contrivance yet evolved for removing sand."

So I wrote twenty-five years ago, and it is still true of Skip and Tip.

OTHER CONTRIVANCES

Back in the old days, when Cornish Plunger poles were plunging up and down, one difficulty was to keep them from airing through the packing. When air got in, the old pitman would say, "She edn havin' solid", or the draught engine driver would say, "I can't get the cataract to leave the handles hardly long enough to fire the boilers". Just as it is necessary to have a "water-seal"

80

for the modern centrifugal, so water around the plunger was necessary to seal and keep out the air. To do this more efficiently, in all states of worn pole and stuffing box, I had cast at Charlestown Foundry boxes and glands with "dished-out" tops, to contain water enough to outlast the tremendous vacuum caused by the instantaneous 9 ft. upstroke of the pole. This greatly helped the smooth working of these old pumps.

HYDRO-VAC

Another gadget very useful in an emergency, where there is a water pressure of at least 100 lbs. per square inch, is what I named the Hydro-Vac, or, as it is usually shortened into, "the Vac". I had experimented with improvised fittings and played around with this idea for years.

In an emergency, I made a sketch and sent to Drinnick Fitting shop for the parts I wanted, and we - my sons and the shift-bosses - assembled it with very successful results, even above our expectations.

This is the story Mr. W. O. Meade-King told at a Captains' meeting, a little while after he was made Head Works Manager of English Clays Lovering Pochin.

He said that he was at a Board meeting presided over by Lord Aberconway, when he told the Board that he would like the firm to have a works bonus scheme, whereby anyone, manager, captain or workman, who brought forward anything beneficial to the company should have some recognition and monetary reward for it.

His Lordship enquired, "Have you anything in mind?"He said, "Yes", and went on, "Capt. M. Arthur of Trethosa has evolved what he calls a "Vac". We had a slip of some hundreds of tons of clay down below sand pit level; he set his gadget to work, continued washing, and with it pumped the sand up into the patent pit, where it could be dealt with normally, without any loss of production. Then Dubbers wanted 500 tons of a kind which was below sand pit level; it was used there, and the market was met on time. At Lee Moor a level was being driven downhill to connect to the main level; it was used there to allow the men to work dry, which otherwise could only have been done by laying down motor and pump."

"Then at Wenford, 700 tons of clay was isolated in a tank, away from the only working kiln, the other kilns being full of Government stores. He, with his Vac, pumped the clay to where it could be used, simply by utilising the water pressure in the pipeline from Stannon Pit."

"In fact," Mr. Meade-King said, "it has been worth hundreds of pounds to the firm."

"What would you like to give him?" said his Lordship.

"At least £25", said my manager.

"Give him £50," said Lord Aberconway, which I thankfully received.

Looking back over the years, now that I'm three-score and ten, I'm glad to feel that at least sometimes I've pulled my weight, and done a little to help forward the van of progress.

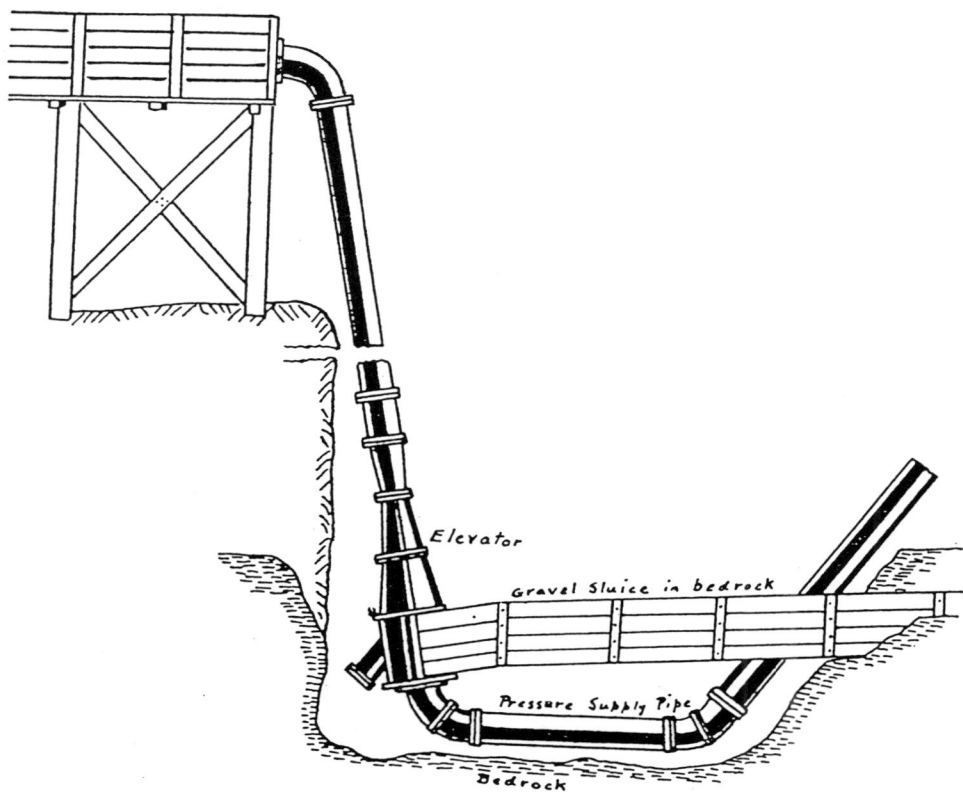

Hydro Vac

Trethosa micas and power house, Kernick pit in background

CHAPTER X

THE CHINA CLAY INDUSTRY

EXTRACTS FROM A LECTURE GIVEN TO THE ST. AUSTELL OLD CORNWALL SOCIETY

"China Clay - a short history of the discovery and rise of the industry, and some of the uses to which the refined clay is put."

History records that china clay, as the name implies, was worked over 2,000 years ago, in China. I have read that it was first obtained from a hill named Kao-ling, from which the commercial name, Kaolin, sometimes used, is taken.

It is said that the secret of the manufacture of porcelain from clay was brought to Europe about the 16th century by a Jesuit missionary, and during the 17th century Italy, France and Germany each became famous for its peculiar china, which is still known by its distinctive mark. During the 18th century, under Josiah Wedgewood, this country came into the picture with its Derby, Crown Derby, Worcester and other famous wares.

So much for pottery; let us have a look at the clay which helps to form the body of china or porcelain.

You probably know that natural china clay is decomposed, or rotten, granite, dug from the earth in Cornwall and Devon, and in other parts of the world. It might help for a moment, if we consider the geological formation of the clay beds. In the very old Book, the Book of Genesis, we read that in the beginning God created the heaven and the earth, and the earth was without form and void. Geologists say that it was an "uncompounded, homogenous, gaseous condition of matter", which, far from making it any plainer than the Bible, makes it as clear as mud to the ordinary man.

If we delve into geology, geologists, like the Bible, give different periods in the formation or solidifying of the Earth's crust - in the Bible it's days, and the geologists call its periods palaeozoic, mesozoic, cainozoic, or primary, secondary and tertiary really three main periods, each sub-divided.

Apparently in the first period, the Cambrian or Devonian china clay districts were formed. At first red sandstone somewhat similar to the cliffs around Torbay overlay everything, until it suffered a granite intrusion. The late Mr. Joseph M. Coon, F.G.S. told of a magma reservoir under Cornwall and South Devon, which erupted like a volcano, forming the Scilly Isles, the Penryn, the St. Austell and the Dartmoor upheavals of granite. It appears feasible, for on going down the old engine shaft at Parkandillick, near St. Dennis (the shaft was sunk in killas, a kind of transformed sandstone), and walking a little way into the "level" leading away from the shaft bottom, you see how the ground abruptly and vertically changes from killas to granite as if it was pushed up from below.

When the granite cooled (for granite is an igneous rock, which means that it has been in the fire, differing from sandstone, which is a sedimentary rock) it shrank and split, leaving crevices. Then the spar lodes were later forced up through from the liquid reservoir underneath, cooling suddenly on striking the cold sides of the rock, and it gave off a hot hydrofluoric acid gas, which permeated and broke down the original granite into what we call virgin clay. At least, something like this happened. I'm not a geologist or chemist, but I know from experience that the best clay is found near lodes - we have a saying, "nearer the bone, sweeter the meat."

86

Here was the clay awaiting discovery. About two hundred years ago a Plymouth chemist, William Cookworthy by name, visited Cornwall, and being interested in pottery as well as plying the trade of a chemist, he first found china clay at Tregonning Hill, in the parish of Breage, near Mounts Bay, then later, in larger quantities, in the parish of St. Stephen-in-Brannel.

Mr. Cookworthy was a native of Kingsbridge, Devon, and probably used Devon ball clay at his pottery, but was always on the look-out for something better. Actually, he visited St. Stephen-in-Brannel in search of china stone, and found both china clay and china stone at Carloggas.

According to what the late Mr. C. E. Davies told me, he stayed with a farmer named Yelland. My great-grandfather, Tom Arthur, married Yelland's daughter, Patience. Though only a tiny link, it interests me to know that my family has been connected with the industries from the start, principally with china stone.

In the St. Stephen Register of Births is recorded that to Thomas and Patience Arthur, of Carloggas Quarry, was born a daughter Dorcas, a family name. I had a sister, and have a niece so named.

At first china stone developed almost as fast as china clay. Dr. Boas, writing in 1830 of our parish in Gilbert's History of Cornwall, after describing the geological formation, says, "But the most abundant variety is that extensively decomposed into a white friable mass. Of this substance, the more compact and perfect parts are quarried and exported under the name of china stone, whilst the earthy parts are washed, to separate the fine argillaceous particles from the grains of quartz, and from the other hard minerals with which they are mixed in their natural state. The water, rendered like milk by this, is received in large pits, where the earthy part subsides, and then, after being dried, is sent to the potteries under the name of china clay. The manipulations which this substance undergoes during its preparation are simple, but very tedious, and they afford employment to several hundred persons. The extent of this branch of commerce is so great, that about 7,000 tons of clay and 5,000 tons of china stone are exported from Cornwall every year, of which by far the greater is raised in this parish."

To go back to Cookworthy, an article in the "Polytechnic Magazine," in the year 1850, says that Cookworthy took samples of clay from Lescrowe and Trethose. The writer who penned this evidently had a peculiar way of making his final "a" and made it like an "e". He must have meant Rescrowsa and Trethosa (old men still call it Lescrowsa).

It was in Carloggas Moor, on the very edge of Rescrowsa Farm, that the first clay pit was opened, afterwards working across the border into it. Trethosa, too, was worked very early in the 19th century. I once saw a map of a clay-sett, taken up in 1804, signed "Camelford". Lord Camelford, who then owned the Boconnoc Estate, was a nephew of the Earl of Chatham; both Carloggas and Trethosa were on his property. Lord Camelford entered into partnership with Cookworthy, and started a pottery at Plymouth (according to Fairlie), removing later to Bristol, but somehow it never became a paying concern, when the monopoly expired apparently the Wedgewoods, who had potteries in Staffordshire, made a go of it.

To return to this district, when the pits first started, a pair of "streamer's boots", a long-handled shovel, a wheelbarrow, and perhaps a hand-pump were what composed the machinery and plant of the clay works. Very probably the tin-streamers (the first wearers of those high leather boots) had found the clay while streaming for tin, and as the clay pioneers first chose the places where there was only shallow overburden, the wheelbarrow and horse and cart preceded the tram and skip-wagons for removing overburden.

There being no monitor-hoses then, clay breakers and water down over the strakes provided the method. The sand was "landed" or, where the lay of the land permitted, washed to a lower level, and the clay ran through an adit level to settle in small cut stone pits (the joints packed with moss), instead of pointed with cement. From there it was run or pumped into shallow sanded pans to solidify, afterwards, in fine weather, carried by men, women and boys, and spread over the ground to be dried by the sun and wind, later stacked in hollow piles, and covered with thatched reeders when it rained. When bone dry, it was scraped clean by women clad in white aprons and gook bonnets, then stacked ready for export.

Vast improvements in the hauling and pumping plants have taken place - in fact they had to! As the pits deepened, the wheelbarrow for dealing with sand would be worse than filling a lorry with a teaspoon, and the hand-pump for pumping from a deep pit as impractical as Tregeagle emptying Dozmary Pool with a broken limpet shell.

The refineries, pits, tanks, kilns, all have been wonderfully improved (I wonder, by the way, where the first drying kiln was built?). Capt. Tom Yelland, who was born in 1828, once told me that it was either North Carloggas or Parkandillick. I wonder if there's a record anywhere.

The growth of the clay and stone trade is shown from a few thousand tons in 1830 to nearly a million tons in 1913, and in 1938 it came up above again. War is a great disturber of trade - so we must learn to outlaw war - and I heard a Clay Manager say recently that there is no reason why it should not rise to one and a quarter million tons per annum, in the not too distant future.

Many people think that clay is used only for crockery, insulators and sanitary ware, for which it is potted and baked, but much more clay is used in paper-making. I've seen paper with from 20 per cent to 75 per cent clay in it; and, of course, clay is used in a host of other things, from hot water bottles to bootsocks.

More than twenty years ago, "Investigator" in the "China Clay Trade Review" wrote that china clay was used for many things. He gave a list, then added that he didn't think it complete, as clay was used for some purposes known only to the users themselves, and he went on, "It's use by the paper, pottery and textile trades is well known, and does not call for detailed reference, but some of the lesser-known uses to which china clay is put are washing and powdery cleansing soaps, water softeners and sewage purifiers, metal and plate cleaners, stove and boot polishes, toilet powders, cosmetics, tooth powders and pastes, ultramarine, alum, starch, chemical manures, and fertilisers, disinfectant, powders and paints, crayons and pencils, linoleums, clay beds for writing and typewriting duplicators, picture-frame moulding, asbestos, fire bricks, boiler packing, plaster, whitewash, modelling materials, buttons, knife and fork handles, papier mache, india-rubber, dance compo., cleaners for white canvas shoes, composition for marking out sports grounds, as a substitute for talc, builders' plaster, sculptors' clay, plaster of Paris and washable distemper. It is also largely used as an adulterant in food, which, for obvious reasons, I will not specify."

This list was printed some twenty-five years ago, and we know many other uses have been found since then, such as kaoline poultices, and I'm told colloidal clay enters largely into the making of stomach powders. It is claimed that the healing properties of these finely-divided clays have a soothing and healing effect on an ulcerated stomach.

From all this, one can say that china clay and stone are very important industries, with a past, present, and future.

Now I wonder if I can, briefly, describe the processes. Most people in mid-Cornwall would know that there is a process. "Foreigners" from the other side of the Tamar don't, always! I've known Londoners exclaim, "Oh! You have to do this, that, and the other. I really thought all you had to do was to chop it out with a spade, put it in sacks, and it was ready for the market."

To begin, a lump of virgin clay has exactly the same ingredients as granite, the only difference being that the clay is soft, instead of hard like the granite. In them there are three main ingredients; though there are small portions and traces of magnesia lime, oxide of iron, vegetable stain, humic acid, etc., the main parts are clay, sand and shell, or felspar, quartz and mica. The clayworker's job is to separate the clay from the others to make a pure uniform substance, commercially known as kaolin or china clay. I think to the chemist it is a hydrated silicate of alumina.

I told a Yorkshire farmer (on holiday) in 1947, that in the clay pits it's threshing day every day, all the year round. One might say that the clay was the wheat, the sand, the straw and the shell, the "doust"; ours, though, was a wet thresh - water was the divider, separator, as well as the conveyor. Where there were formerly men with "hackers and dubbers", a pressure hose is now trained on the raw clay, which is taken up by the water or the shell, the sand stays in the pit by the simple method of having a small pool in front of the trap launder which runs off the clay stream.

Thus the sand is eliminated. It is from four to five times as much as the clay, and after draining it is hauled up by the skip-wagon, to make the conspicuous conical mounds so familiar to mid-Cornwall folk. Let us follow the stream which now goes to the pump, and is lifted to the "round heads" at the beginning of the refineries, locally called micas, where it is spread out and run through shallow, narrow channels at a controlled speed, to give the clay and shell a chance to sort themselves out.

Here again, gravity comes into play, because the larger particles are heavier than the finer ones; they drop to the bottom, while the clay goes forward to the settling pits. These channels are periodically cleaned with a tool called a "shiver", by pulling out plugs and pushing the shell through the hole and on to the mica-dam.

To go on to the clay. This is left long enough in the settling pits to thicken to the right consistency for "landing" to the kiln tanks or press sumps alongside the kiln, and sometimes miles away from the pit. It is then trammed on the kiln pan (old method) or pumped through the filter press, and the cakes finished off on the kiln pan floor. This method takes less coal, but is not all that cheaper, as press sheets and plates are very dear, but it's much speedier.

There are also modern rotary and "buel" driers. The Rotary is something like a huge Cornish or Lancashire boiler, with a fire in it slowly rotating and being tilted up at the receiving end. The pressed clay, cut into strips, goes in one end wet, and comes out the other end dry, to drop on the conveyor belt, and away to the storage linhay. The Buel is vertical instead of horizontal, and steam is the drying agent. A disintegrating machine is used sometimes with this method, and the "dis" clay drops straight into bags, being automatically weighed and sealed, ready for the market. You will see that this is a far cry from the days when all clays were sun-dried.

It would take too long to trace the evolution of the pumping and hauling plants through the years, although that is a fascinating study to anyone interested in machinery. However, I hope by this simple paper to have given some idea of the extent of the improvements already achieved and, above all, of the importance of our staple industry - china clay and stone.

Finally, just a word about the workers. Being Cornish, probably I'm prejudiced in their favour, but for ability and reliability you will have to go far to find their equal. I speak not without some knowledge of men from different parts of the world, having had under me Kaffirs, Arabs, Chinese, Germans, Australians, Poles, Jews, Austrians, New Zealanders, Newfoundlanders, Canadians, Americans, French, Flemish, Scottish, Irish, Welsh, and from many counties of England. I found good, bad and indifferent among them all.

But for standing up to hard continuous labour, the clay-workers top the list. Consider the enormous amount of sand, in a shift, that a couple will fill, tram and dump into a skip-wagon - often 20 fathoms (about 200 tons). They are sturdy men: they need to be, with the exposure to wind and weather in the open clay pits, three shifts about.

One jocularly remarked, "It's vitty grand - in the summer like a sanatorium, in the winter all the same as a refrigerator, so we are well preserved all the year around."

Yes, their sense of humour, the ability to joke, and their equally strong sense of duty, carry them through. Generous out of their small wage-packet of pre-war years, always willing to help, they have given hundreds of pounds to sick or injured comrades, and to hospitals and deserving causes.

Then, among them, local or lay preachers, living the life weekdays, and before transport was like it is today - walking scores of miles on Sundays to keep the religious life of the community going. Not all saints, I'm not saying that, but individualists, independent, witty.

Take Dick Hi-up, a clay worker I knew nearly sixty years ago. That, of course, was his nickname, for riding along on his Goss Moor pony with a greyhound dog alongside, he would often shout "Hi-up, there!" Dick liked a drop of beer, and when hard-up sometimes got the wherewithal as follows.

Some of you will remember the "three life system", where a man took up, say, 20 acres of common land, broke and brought it under cultivation, built hedges and also a farmhouse, on the condition that while he, his wife, and often his eldest son, lived, he should enjoy it for a nominal rent. Dick's father had done this, but when Dick inherited it he sold the lease to have some ready money, which soon melted.

Now for an example of his wit. When really hard up, he would run through the farmyard shouting at the top of his voice, "Life is not worth living! I'm going to drown myself!" Out would come the smallholder's wife, saying, "Whatever is the matter Richard?" "Oh, I asked a man to lend me half-a-crown and he wouldn't do it. I tell you, without a drop of beer, life isn't worth living, so I'm going to yonder pond and end it all!" "No, no, Richard, you mustn't do that; I'll give you that half-a-crown rather than that." You see, he was worth more alive than dead to them. On his death, house and land in good repair would revert again to the landowner, so Dick, the wily rascal, got his drink.

When in funds after the monthly pay-day at the clayworks, Dick sometimes got a trifle "o'er the Bay" as sailors say, and two or three times a year got called before his "betters". On one occasion the Bench chairman (who had a squeaky, effeminate voice) said, "Hallo! You here again! 5s. and costs!" Dick said, "Speak up, Sir Charles, you're like a man shouting out under a knife!" "10s. and costs," said Sir Charles. "Oh! Oh! I can hear that all right," said Dick. That was one time, anyway when his wit wouldn't work.

Then we had our local Dr. Spooner, you know, the Professor who somehow got a lot of his words inverted, or forth and back, such as when describing his luggage as two bugs and a rag, or at a wedding saying "It's kistamary to cuss the bride". Well, old Joe was much like that, only more so. This is how he described a storm when the Captain asked him what sort of weather it had been during the night.

"Oh, Cappen, a vitty awful night. I zeed thunder, heerd lightening, pulled my eyes down over my hat, and ran down the incline as fast as I could walk!"

On another occasion, after being at an auction sale, he said, "I saw a beautiful mahogany table sold: it was all oak, and not a skerrick of deal in it!" And so old Joe provided endless amusement among his comrades.

Another humorist was Dicky. He had a peculiar husky voice, difficult to describe, and a habit of winking or shutting both eyes at once. One story was of how he raced with a gypsy home from Summercourt Fair. He was proud of his horse, but had rather a ricketty trap. Going along trotting, a gypsy passed him at a gallop. "So, it's a race you want, is sit?" said Dicky. So he opened out, and got alongside the gypsy; but the rough road was too much for his trap going at full gallop and the wheel spokes began to fly up around his ears. He shouted to the Gypsy (not knowing where they came from), "If you're going to race, race fair, don't go shlowing shlicks "(throwing sticks). To hear him go on in his inimitable way made fun for us all, young and old.

There were also a few humorous scoffers at religion, such as Charley B. He said, "I went through Foxhole t'other evening, and there was Captains T and B singing at the top of their voices, "We are out on the Ocean sailing." Biggest lie I ever heard in my life" says Charley, "they weren't out on no Ocean 'tall, but in Foxhole Chapel." Then take his description of a Prayer meeting. He says to me, "What do you think, Cappen? I was at a Prayer meeting last evening, and a man was praying ever so earnestly, "I thank Thee Lord that I can worship Thee under my own vine and fig tree" and the man never had a flatpoll cabbage in his garden, let alone a vine or fig tree!" Humorists, not infidels, indulging in a little Methodist leg-pulling.

Then remedies for illness. Faith cure for brow ague: write your name on a slip of paper, but if there are t's or i's in your name you must not cross a "t" or dot an "i", that would be clean against the charm, and spoil it all. When your name is on the paper, cut a slit in the bark of an elder tree, then the pain disappears like magic. Someone told Phil Longstone (a boozer) this story. He said, "I verily believe it, for once I got drunk on elderberry wine, and had a fearful headache next day. The pain must have been in the tree."

Such were some of the characters, and the stories they told, but in the main they were honest, reliable, efficient, capable, and a hard-working community, who, through the years, under wise management, built up our local industry - men whom I have been, and still am, proud to have known and with whom I am happy to have laboured.

A TRIBUTE TO CAPTAIN MARSHEL ARTHUR

The china clay industry was the poorer by the death at Foxhole on Friday, 23rd February, 1962 of one of its most well-known and well-loved grand old men, Captain Marshel Arthur. He was within two or three months of his 83rd birthday.

Bard, writer, broadcaster, but first and foremost a china clay man, Cap'n Marsh Arthur, as he was known throughout the Duchy, was one of the best known clay captains. And in his own long lifetime he became almost a legendary figure.

Born on the 2nd May, 1879 at Middle Hill, Foxhole, he was one of thirteen children. When he was only five years old his father, a china stone quarryman, died and to keep the home together his mother went out to work scraping clay. Young Marsh started school in 1885 at Nanpean but left when he was nine to work for his brother William at Wheal Arthur Quarry.

In spite of only a brief time at school, or perhaps because of it, Marsh studied hard at what, in those days were called "Continuation Classes". And did this so successfully that in 1910 he was a teacher at Nanpean Night School, teaching mathematics, hydraulics and kindred subjects. Indeed, some of his former pupils later held positions of responsibility in the clay industry and overseas.

After a time with his brother, Marsh left to work for Capt. Tom Bray at South Carloggas Clay pit. For many years he was a pit-man for the West of England and Great Beam China Clay Co., and his early wanderings through the industry read like a record of old workings. By 1909 he was Captain at North Carloggas and when World War I broke out he volunteered, serving in Egypt and France, during which time he was awarded the Military Medal.

With demobilisation Cap'n Marsh returned to North Carloggas, transferring to Trethosa in January 1925. He retired in September 1947, and his long years of service provide a saga of hard work, initiative and achievement. Then projects in which he played a notable part or had a hand in are legion: Trethosa Micas, a patent sand trap to replace the old ground pit; the installation of the derrick sky tip and the V-shaped skip (with his brother Jim): the Hydro-Vac (with his two sons, Launce and Ralph, and shift bosses, P Bazley and W Biscombe): the "Sky-high" filter for sampling; and other innovations too numerous to mention.

Cap'n Marsh's religious life centred round Foxhole Methodist Church - from scholar, through Sunday School teaching, to Superintendent, Trustee and local preacher. He was a deeply religious man and from an early age was a teetotaller and non-smoker. He loved music and for many years played in the Foxhole Brass Band and achieved some success in festivals.

In 1938, Cap'n Marsh, with the late Mr. Alf Davis made a broadcast on China Clay and, during World War II broadcast to our forces overseas. For 8 years he served on the St Austell Rural District Council, only retiring from office in 1958. He was made a Cornish Bard for his written works on china clay ("Scryfer an pry gwyn") and his articles have been much published in the "Cornish Magazine". He took a great interest in the Darby and Joan Club, delighting them with his never-ending fund of stories and recitations. In all, he enjoyed a happy and active retirement and was a familiar figure in his car, ECV 72, which he claimed stood for

Final three words - "English Clays Veteran".